Sustainability Ethics and Sustainability Research

Christian U. Becker

Sustainability Ethics and Sustainability Research

Springer

Christian U. Becker
Department of Philosophy
The Pennsylvania State University
240 Sparks Building
University Park, PA 16802
USA
cub20@psu.edu

ISBN 978-94-007-2284-2 e-ISBN 978-94-007-2285-9
DOI 10.1007/978-94-007-2285-9
Springer Dordrecht Heidelberg London New York

Library of Congress Control Number: 2011938688

Printed on acid-free paper

Springer is part of Springer Science+Business Media (www.springer.com)

To Katja

Preface

The idea for this book has evolved since 2003 as a result of my involvement in research projects on sustainability issues and teaching of sustainability topics. I have recognized that the ethical aspects of sustainability are often neglected, misinterpreted, or misrepresented, and that there is a need for adequate ways to approach and include the ethical dimension of sustainability in public and academic discussions. With this book, I provide a detailed discussion of the ethical dimension of sustainability and the implications of this dimension for academic approaches to sustainability issues. I hope that this book will support future research and discourses about sustainability. Sustainability remains an important concept for public and academic discussions of the future development of societies. However, the full potential of the concept can be fulfilled only if we recognize and adequately address its entire meaning, and particularly its inherent ethical dimension.

I would like to thank all persons who have supported this book project. In particular, I thank Wolfgang Neuser, who supported my philosophical work since 2005 at his chair of philosophy at Kaiserslautern University, Germany. This book is a revised version of my Habilitation thesis in philosophy which was accepted at Kaiserslautern University in 2009. I am grateful to Wolfgang Neuser, Malte Faber, Olaf Breidbach, and three anonymous referees for their detailed and thoughtful reports on the manuscript. For fruitful discussions and comments during the development of the manuscript I thank the faculty of the philosophy department at Dalhousie University, Halifax, Canada, which I joined as visiting assistant professor in 2007–2008, my colleagues in the faculty of social sciences at Kaiserslautern University, and my colleagues in the philosophy department of the Pennsylvania State University. I particularly thank Lauren Nuckols for her detailed comments and corrections to the manuscript, and Joy Drohan for assistance with the final manuscript preparation.

Finally, I thank my wife and my children for their patience and support, which were the most substantial factors making this book possible.

Christian U. Becker

Contents

List of Figures

Chapter 1
Introduction

Sustainability has become a major topic within the last 25 years. It has prominently been highlighted in important global political documents such as the Brundtland Report (WCED 1987), the Rio Declaration (UN 1992a), the Agenda 21 (UN 1992b), and the Johannesburg Declaration (UN 2002), and has been recognized and discussed widely in public and scientific discourses. Sustainability is a global concept that is used to discuss various societal fields, such as business or education, and to discuss a range of crucial environmental, societal, and global issues, such as biodiversity loss, climate change, distribution and use of nonrenewable resources, energy production and use, global equity and justice, and economic issues.

However, although sustainability is widely acknowledged and discussed as an important topic, there is a tendency to vagueness and unclearness in sustainability discussions. There seems to be no precise meaning of the concept, but rather a large variety of different (often implicit) meanings and usages of the term. One might conclude, therefore, that sustainability is a mere buzzword that everyone uses in a different meaning in different contexts. One may wonder whether sustainability is a fruitful concept for addressing environmental, societal, and global issues—whether this concept helps the analysis and solution of these issues and their further clarification.

I argue in this book that sustainability is an important and fruitful concept, but that its potential needs to be fully recognized and developed. The concept of sustainability will be fruitful only for the further discussion, understanding, and overcoming of societal and global issues if the full meaning of the concept is adequately recognized and considered. It is particularly important to understand that sustainability is a normative and evaluative concept. Sustainability has an inherent ethical dimension and denotes a fundamental ethical issue. This ethical dimension of sustainability, which is often neglected or misunderstood, is crucial to the meaning of the modern concept of sustainability, and needs to be adequately identified and considered in public and academic discussions. I will argue that the ethical issue of sustainability is at its core an issue of the self-identity of the modern individual and the related design of modern societies. Sustainability asks for a paradigm shift, for a replacement of the established ideal of the human being as an

C.U. Becker, *Sustainability Ethics and Sustainability Research*,
DOI 10.1007/978-94-007-2285-9_1,

autonomous and independent individual with a new ideal of the human being as a fundamentally dependent and related being—fundamentally related to contemporaries, future generations, and nature. This normative demand inherent to the modern concept of sustainability must be ethically construed, and its implication for public discourses and sustainability research must be discussed.

This book analyzes the ethical dimension of sustainability and its implications in detail. With this, I aim to clarify the concept of sustainability and to develop adequate ways to discuss and approach sustainability issues. The first focus of this book is on the ethical dimension of sustainability. I argue that without its adequate recognition and analysis the concept of sustainability is misunderstood and sustainability issues cannot be adequately approached, and I provide a detailed philosophical analysis of the ethics of sustainability. Against the background of the results of my ethical analysis I discuss implications for the methodological question of how to design an academic approach to sustainability, particularly of how to integrate ethics and science into a comprehensive academic approach to sustainability issues. The second focus of the book is on these methodological considerations on sustainability research. Both my ethical and methodological results demonstrate the important role of philosophy for the analysis and solution of crucial current societal and global issues. Philosophy allows for the analysis of general structures, characteristics, and causes of these issues; for the identification and analysis of ethical aspects; and for methodological considerations on how to approach them. By this, philosophy can make an important contribution to the academic research on sustainability and, ultimately, to the development of adequate political strategies and action.

With this book I provide a new description of the crucial aspects and questions in regard to sustainability and suggest an encompassing analytical framework for analyzing sustainability issues. I identify the inherent ethical dimension of sustainability and demonstrate that it is a complex and new ethical challenge that requires a specific approach to adequately discuss it. Based on an analysis of the characteristics of the ethical dimension of sustainability, I develop a new approach of sustainability ethics and apply it for a detailed discussion of the ethical aspects of sustainability. The results of my ethical analysis allow for important conclusions on the overall role of academics for the analysis and solution of sustainability issues and for a proper design of an adequate academic approach. I argue that we ultimately need a new type of self-reflective inter- and transdisciplinary sustainability research that integrates ethics to adequately approach the issue of sustainability as a whole, and I give main guidelines for the design of such a new type of sustainability research.

The ethical dimension of sustainability is not an add-on, but rather an inherent component of the concept of sustainability. The ethical dimension is grounded in the specific relational meaning of the term sustainability, which refers to fundamental relationships of the human being: its relations to other contemporaries, future generations, and nature. These relations I call *sustainability relations*. An encompassing academic discussion of sustainability requires an adequate inclusion of the ethical aspects of these relations. I analyze in detail the sustainability relations and their crucial meaning for the issue of sustainability and argue that an encompassing sustainability ethics must be able to simultaneously address all three relations in an integrated way. However, established ethical approaches, such as utilitarianism or deontology, are not appropriate to fully capture all the specifics of these relationships for two reasons: First, they have been mainly developed to address the relationship between contemporaries. They are not originally designed for an analysis of the ethical aspects of the relationship with future generations or with nature. As I will argue, their extension to these relationships raises several problems. Second, the ethical dimension of sustainability has a fundamental systemic aspect. Social and global systems play a crucial role for the constitution and actualization of the sustainability relations. Sustainability ethics, therefore, is not just about fundamental relationships, but is also about an adequate design of societal and global systems that allows an ideal realization of the sustainability relations.

However, the relevant systems that are crucial with respect to sustainability are rather complex patterns of action and thought. I introduce the concept of *meta-structures* to identify and analyze these patterns in detail. Meta-structures are complex compositions of ideas and institutions that mutually enforce each other. Crucial examples are science, technology, and the economy, which are at the center of my analysis. The meta-structures are based on some fundamental assumptions and values, realized by various institutions and organizations, and driven by certain incentives and motivations. Together, these elements of the meta-structure cause its dynamics and persistence. As individuals, we are already located with our thought and actions in the meta-structures. This has a crucial impact on individual thinking and acting, particularly on the way the individual is related to others. Meta-structures influence our self-identity and set us automatically into specific relationships with other humans, future generations, and nature. Therefore, the influence of meta-structures on these relations is an important issue for the analysis of sustainability and its ethical dimension. The question of what would be a good design of these structures in the light of sustainability becomes crucial for sustainability ethics.

Based on the analysis of the specific characteristics of the ethical dimension of sustainability, I provide a conception of sustainability ethics that refers to the relational aspect of sustainability. This conception addresses the two ethical elements that are crucial for the setup of the sustainability relations: individual morality and systemic design. On the one hand, I refer to the self-identity and individual morality of the individual person as a relational and dependent being existing within the three sustainability relations and develop the concept of a *sustainable person*. On the other hand, I deduce guidelines for a proper design of the meta-structures, which impact the individual and its relationships and mediate and govern the sustainability relations. Theoretically, the first concept of individual morality of the sustainable person is used to deduce the guidelines for the systemic level. I argue that the meta-structures must be designed in reference to and coherence with the sustainable person. In practice, however, the ethics of sustainability can be thought of only as a dynamic process of mutual development of individual morality and ethically appropriate systems. It is a process of mutual enforcement and response between individual morality and ethical structures for the sustainability relations.

From the ethical analysis and its results, I draw conclusions for the appropriate way to approach the issue of sustainability on the academic level. I argue that traditional sciences by themselves are not sufficient for addressing the issue of sustainability. This is mainly for two reasons. First, sustainability has an inherent ethical dimension that cannot be analyzed by traditional sciences being restricted to factual analysis, but requires philosophical and ethical analysis. Second, because science itself is a crucial meta-structure, science is not a neutral tool for the analysis of sustainability issues, but is also part of the ethical problem of sustainability. We must analyze the ethical role science itself plays in regard to sustainability, particularly inherent normative assumptions of scientific approaches, and discuss implications for sustainability research. A restriction to the factual dimension of sustainability and to traditional scientific analysis would be inadequate to the subject matter of sustainability. The same holds true for a simple division of labor between science and ethics, in which the former solely focuses on the factual dimension and the latter on the normative dimension of sustainability. We rather need a new type of sustainability research that integrates an analysis of the factual and normative level—science and ethics—and is able to reflect on the inherent ethical role of science as both a tool of analysis and a crucial part of the subject matter itself. I outline main requirements for such a new type of sustainability research in Part IV.

The philosophical approach of this book is characterized by the following methodological characteristics and theoretical philosophical background. First, the book is a work of applied philosophy. It understands philosophy as a method of reflection that allows us to analyze crucial issues of societies and helps to understand, approach, and solve them. Second, my argumentation refers to an analysis of the concept of sustainability in order to determine the meaning of the term and to define its main characteristics. This is not an analysis of the conceptual history of the term, but rather refers to the current use of the term within politics, public discourse, and science. I hold that there exists a modern concept of sustainability that has emerged in the political discourse since the 1980s, and analyze the main characteristics of this modern concept and its meaning. Third, my ethical approach may roughly be characterized as a combination of virtue ethics and critical theory. This is, however, no explicit reference to specific philosophical works or schools, but rather a reference to the general characteristics and ideas of these theories. In regard to virtue ethics, I refer to the general idea that human excellence and virtues are crucial ethical themes and that virtues are determined and developed in the context of certain (social) relationships. If we define critical theory in a general way as a theory that analyzes the preconditions of thought and action given by social structures, one may say that this book is related to this tradition as well, insofar as I put particular emphasis on the ethical meaning of meta-structures. Fourth, my methodological approach to sustainability research is not a descriptive philosophy of science, but a normative methodology. It is based on the assumption that guidelines for an adequate approach can be deduced from the characteristics of the respective subject matter. In the case of sustainability, the subject matter particularly shows crucial ethical aspects that imply specific requirements for a sustainability research.

This book focuses on the analyses of sustainability issues and related ethical and methodological aspects. I emphasize the philosophical contribution to the understanding and academic analysis of the current topic of sustainability. As a philosophical work, this book intends to reach philosophers who are working in the areas of ethics and applied ethics, environmental philosophy, or philosophy of science and technology, or who are generally interested in the role philosophy can play in the analysis and solution of sustainability issues. However, the book is also written for a broader audience and aims to reach scholars, students, politicians, and others who deal with sustainability issues and have a particular interest in ethical and methodological aspects. I particularly aim to foster a dialogue among researchers and philosophers about the development of a new type of sustainability research.

References

United Nations (UN). (1992a). *Rio declaration on environment and development*. New York: United Nations.

United Nations (UN). (1992b). *Agenda 21*. New York: United Nations.

United Nations (UN). (2002). Johannesburg declaration on sustainable development. In United Nations, *Report of the world summit on sustainable development* (pp. 1–5). New York: United Nations.

World Commission on Environment and Development (WCED). (1987). *Our common future*. Oxford: Oxford University Press.

Part I
Sustainability and Ethics

Chapter 2
The Meaning of Sustainability

Sustainability has become a crucial concept in global public and political discussions in recent years and also has increasingly entered into the scientific discourse. As a single word, sustainability seems to express some of the main future challenges of humanity and also seems to represent a major reference point for human decisions and activities in the long run. It is, however, difficult to determine the exact meaning and to provide one single definition of the term *sustainability*. The term is used in a broad range of meanings in public and scientific discussions. One might even consider the term to be a useless buzzword, as it seems to encompass rather different and sometimes even seemingly incompatible meanings. However, I hold that sustainability is more than just a stylish catchword and that a closer look at its usage in political and scientific discussions since the 1980s reveals a specific basic meaning. From a philosophical perspective one can identify three main characteristics that determine the core meaning of the modern concept of sustainability and the fundamental issues to which it refers.

1. The meaning of ***continuance***. Literally, the term sustainability means the ability to keep going, to keep up, to maintain, to cause to continue in a certain state.[1] Sustainability refers to the continued existence of something over time.[2] When using this concept, we may refer to a system (e.g., an ecosystem, an economic system), a certain entity (e.g., a species, building, capital), or a process (e.g., evolution, an activity). Further, the term sustainability allows for two interpretations: it can be understood as the ability of a system, entity, or process to maintain itself, or the ability of humans to maintain a certain system, entity, or process. Examples fitting to the first interpretation are ecosystems, species, or biological evolution. Examples of the latter case would be economically used ecosystems, such as forests, or grazing management. With its basic meaning of continuance,

[1] See, e.g., *The Oxford English Dictionary* (Simpson and Weiner 1989).

[2] Some definitions of sustainability are based on the aspect of continuance, e.g., the definition given in the *Shorter Routledge Encyclopedia of Philosophy* (Holland 2005).

DOI 10.1007/978-94-007-2285-9_2,

the term sustainability refers to the idea of some kind of stability in time. It is the idea of distinguishing stable factors in a context of dynamics and change.

Continuance is the aspect of sustainability to which science traditionally refers. The scientific discourse frames sustainability issues in various ways as issues of continuance. In approaching sustainability, science usually defines certain characteristics of specific systems, processes, or entities to discuss the continuance of these systems, processes, or entities. In this perspective, for instance, "a sustainable ecosystem is one that, over the normal cycle of disturbance events, maintains its characteristic diversity of major functional groups, productivity, and rates of biochemical cycling" (Chapin et al. 1996: 1016). Scientific approaches to sustainability differ depending on the way a system, entity, or process is defined, the time horizon is chosen, and the parameters of study are determined. Approaches of natural sciences typically analyze the capability of certain natural systems, entities, or processes to continue—that is, they determine and analyze respective parameters and conditions of its dynamics. In this perspective, the ability of humans to sustain a system is the ability to keep or improve the relevant parameters of continuance of the system (see, e.g., Chapin et al. 1996). Within an integrated analyses by natural and social sciences together, human decisions and actions can also be considered as part of the system defined. The relevant parameter of the continuance of such a social-ecological system could be income, capital stock, or welfare.[3]

The differences between scientific definitions of, and approaches to, sustainability result from the different subject matters of scientific disciplines, which make them focus on different systems, entities, or processes. Differences between scientific approaches further result from specific choices of assumptions, parameters, and time horizons underlying specific research approaches. However, despite these differences all scientific approaches have in common that they focus on the continuance aspect of sustainability. They provide us with analyses and explanations of the dynamics of various systems, processes, or entities, and formulate conditions for keeping them up.

With the focus on continuance, science certainly has its role in the discussion of sustainability, because it addresses one core meaning of sustainability. However, the scientific understanding of sustainability as ability for continuance covers only *one* aspect of the meaning of the modern term sustainability and, therefore, even the whole spectrum of scientific contributions to sustainability addresses only a part of the sustainability issue and is not sufficient for an encompassing analysis. To understand this limitation and consider the implications for academics to address sustainability, one has to recognize that sustainability entails two further core meanings in addition to the meaning of continuance.

2. The meaning of ***orientation***. The modern use and understanding of the term sustainability shows an inherent normative and evaluative meaning, and it would be a misinterpretation and misrepresentation to disregard this meaning in academic

[3] See, for instance, Berg and Hofkes (1999) or Quaas et al. (2007) for possible ways to analyze the continuance of social-ecological systems.

discussions.[4] Today, sustainability is widely used as a norm. Sustainability is regarded as being something positive, something for which we should strive. It is seen as a major aim and an orientation of long-term human actions. The main political statements about sustainability clearly demonstrate this meaning of orientation. For instance, the crucial United Nations international agreement *Agenda 21* states that "sustainable development should become a priority item on the agenda of the international community" (UN 1992: § 2.1), and sustainability is included among the fundamental guiding principles of the international community (*United Nations Millennium Declaration*, UN 2000: §§ 6 and 22f).

The simultaneity of the continuance aspect and the orientation aspect in the meaning of the modern term of sustainability has caused problems and led to confusion. On the one hand, environmentalists and scientists tend to—often implicitly—mix both aspects. They deduce normative aspects from positive aspects and try to give arguments for the orientation aspect—that is, that we should sustain something—based on their scientific insights into the mechanisms of its continuance.[5] This is, however, in contrast to the self-identity of modern science as an endeavor focusing on facts, and it also may be considered a logical fallacy in regard to the is–ought distinction.[6] The normative meaning of sustainability cannot be deduced directly from its meaning of continuance, and it cannot be dealt with within a traditional scientific approach. On the other hand, therefore, many scientists have been skeptical about the normative aspect, and some would like to exclude this meaning entirely from an academic analysis.

However, to restrict the academic discussion of sustainability to traditional scientific analyses of the continuance aspect would mean that academics could say nothing to an important immanent dimension of sustainability, that is, to its normative dimension. Moreover, as I will demonstrate in my further analysis of the normative dimension of sustainability, this would even mean that the contribution of academics to the discussion of sustainability issues would become misleading and counterproductive. I will argue that science itself cannot be considered a neutral, value-free tool in regard to sustainability, but entails normative assumptions and ethically relevant characteristics, by which science becomes also part of the sustainability issue and its normative dimension. Therefore, a simple separation of scientific analysis from normative discussions is not feasible in regard to sustainability.

---

[4]That sustainability has a normative dimension has already been recognized, e.g., by Becker (1997), Newton (2003), Ott and Thapa (2003), Norton (2005), and Clark et al. (2004). However, this has not led to a separate field of sustainability ethics so far.

[5]This has, for example, prominently been the case with Aldo Leopold (1887–1948), one of the most prominent precursors of environmentalism in the twentieth century. His arguments for a *Land Ethic* are based on insights of ecology in his times and result in this conclusion: "A thing is right when it tends to preserve the integrity, stability, and beauty of the biotic community. It is wrong when it tends otherwise." (Leopold [1949/53]1966: 262).

[6]The is–ought distinction has prominently been addressed by Hume ([1740]2000: 3.1.1.27) and, although it raises many philosophical problems, is still one of the fundamentals of modern scientific self-identity. This extends also to the social sciences (see prominently Weber [1918]1988).

A comprehensive and appropriate academic approach to sustainability issues requires consideration of the idea and meaning of sustainability as a whole, and discussion of the possibilities for academics to include the orientational meaning, that is, the normative and evaluative aspects of sustainability, into an overall academic approach. However, before proceeding with the discussion of such an approach, it is important to consider a third core meaning of the term sustainability, its relational meaning, which is essential for the understanding and further interpretation of the orientational meaning of sustainability.

3. Sustainability is about fundamental ***relationships***. That relationships are crucial for the modern meaning of sustainability becomes immediately obvious when we refer to the most common and most accepted definition of sustainable development, given by the Brundtland report *Our Common Future* (WCED 1987). Here, sustainable development is defined as "development that meets the needs of the present without compromising the ability of future generations to meet their own needs" (WCED 1987: Ch. 2, 1). The most important aspect of this statement is the reference to two fundamental relationships of human beings: The relationship between humans and their contemporaries—that is, between different individuals and groups within the present generation—and, secondly, the relationship between the present generation and future generations. The Brundtland definition has become crucial for the further public, political, and scientific use of the modern term sustainability, and it is important to understand that sustainability therefore is not merely about the continuance of something, but also about certain fundamental relationships of human beings (see also UN 1992: 3.33; UN 2002: Annex II, pp. 154–62; Clark 2003: 1–2). In addition to the relationship between contemporaries and the relationship with future generations, there is a third relationship involved with the modern idea of sustainability: the relationship between humans and nature. This relationship is addressed both indirectly and directly.

Indirectly, one refers to the human–nature relation when addressing the relationship among contemporaries and with future generations. Both of the latter relationships are heavily influenced by environmental actions. Particularly, our relationship with future generations is to a large extent an indirect and asymmetrical one, mediated by the long-term effects of our environmental actions and resulting environmental changes. Thus, when we address the relationship with future generations with the modern concept of sustainability, the relationship between humans and nature plays a crucial role. This has particularly been recognized and expressed in a variety of major international political statements on sustainability (WCED 1987: Ch. 2 and 13; UN 1992, 2002: Annex II, pp. 154f).

Moreover, the relationship between humans and nature is also directly addressed by the modern concept sustainability. Sustainability is also about the self-maintaining ability of nature—ecosystems or evolutionary processes—and the human impact on that ability. This is not merely discussed in regard to the consequences of that impact for other humans or future generations, but also in regard to the consequences for nature itself, although this distinction is often not made strictly and explicitly. We can find the direct reference to the human–nature relation in crucial political statements

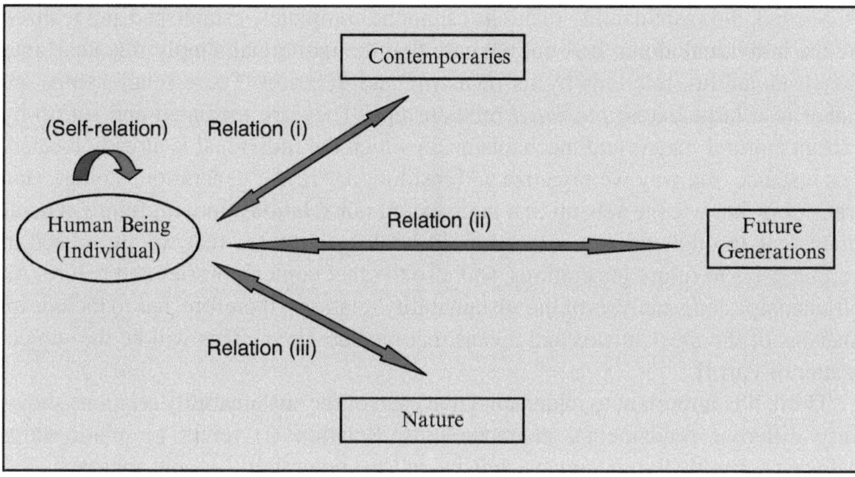

Fig. 2.1 Sustainability relations

such as the *Johannesburg Declaration* (UN 2002: Annex II, pp. 154f) or the Brundtland report, which, for example, states: "In its broadest sense, the strategy for sustainable development aims to promote harmony among human beings and between humanity and nature" (WCED 1987: Chaps. 2 and 81). The human–nature relation is also at the core of many approaches of the natural sciences that analyze the sustainability, that is, the continuance, of certain natural systems as an aspect of itself, and consider human impacts on such systems as externally given factors.

Thus, the modern term sustainability, by its very meaning, refers overall to three fundamental relationships of the human being which I, in short, call *sustainability relations* (Fig. 2.1):

(i) The relationship between humans and their contemporaries
(ii) The relationship between currently living humans and future generations
(iii) The relationship between humans and nature

It is important to recognize here three basic characteristics of this threefold set of relationships, which are crucial for the further analysis and will be discussed in detail later in this book.

First, the sustainability relations have factual *and* normative aspects. We may identify several factual characteristics that describe how we are related to other humans and nature. Relations (i) and (ii) are, for instance, the subject matter of several scientific disciplines such as economics, politics, and sociology. In regard to relation (iii), biology, genetics, and ecology, for instance, analyze biological mechanisms that define our relation with nature. However, there are also normative aspects concerning the three sustainability relations. We can understand, design, and actualize these relations in different ways, and we can ask how we should do so. How ought we relate to contemporaries, future generations, and nature? What would be the best way to understand, design, and actualize the sustainability relations?

Second, the sustainability relations cannot be completely established and realized by the individual alone. It is not the case that the individual simply sets itself into the sustainability relations by its own way and decision. These relationships are rather to a large extent *mediated* relationships. They are mediated and set up by certain cultural means and mechanisms, in which the individual is already located. For instance, the way we organize and institutionalize the generation, storage, and transfer of knowledge sets up to a large extent our relationship with future generations. It is through science, education, technology, art, etc., that we are related in certain ways to future generations, and also to other contemporaries and nature. An all-encompassing analysis of the sustainability relations, therefore, has to include an analysis of the mechanisms and means that mediate them. This will be the subject matter of Part II.

Third, it is important to understand that each of the sustainability relations shows very different fundamental characteristics. Relation (i) refers to relationships between currently living humans, which—at least potentially—could approach each other, recognize each other, and communicate with each other. Relation (ii) shows an abstraction and asymmetry, in so far as these options of direct approach are not fully given—at least if we consider generations of the more distant future. We could merely communicate with them in one way, for example, by bequeath messages, knowledge, artifacts, etc.; and we might try to imagine some of their thoughts, feelings, or values. Relation (iii) is even more difficult. For instance, a communication with non-human nature, if it is possible at all, differs much from a communication between humans. The characteristics and specifics of each relationship and their ethical implications will be analyzed in detail in Part III.

The sustainability relations, therefore, are a complex issue. A comprehensive sustainable research has to be able to analyze all of them in an integrated way. This is a larger challenge for two main reasons. First, modern science is highly separated into disciplines, of which each is restricted to one specific factual part of one of the relations. Each scientific discipline approaches different positive aspects of a specific relation with different theories and methods. An encompassing approach to all three relations would require an integration of all the different disciplines that are of relevance here. The second reason is the difficulty of an integration of the analysis of positive and normative aspects, which traditional science can not deal with. The challenge and potential ways of such integration will be discussed in Part IV.

In summary of the discussion of the sustainability concept, the core meaning of the modern concept of sustainability encompasses three aspects: continuance, orientation, and relationship. To express this by one definition, one may say that *sustainability is the ability to establish continuance as a means for orienting human actions and life toward the threefold relatedness of human existence to contemporaries, future generations, and nature.* In other words, sustainability addresses our ability to recognize and realize ourselves as fundamentally relational beings, as beings embedded in the threefold relationship with others, future generations, and nature. It addresses the human being as a timely, socially, and naturally contingent being and the implications of this threefold contingence for human self-identity, life, and actions.

The adequate recognition of the full meaning of sustainability will allow us to discuss its ethical dimension and develop an adequate ethical theory of and approach toward sustainability. In the following chapter, I will proceed to construe the ethical dimension of sustainability as inherently determined in the meaning of the concept of sustainability.

References

Becker, B. (1997). Sustainability assessment. A review of values, concepts, and methodological approaches. *Issues in agriculture 10.* Washington: The Consultative Group on International Agricultural Research, World Bank.

Chapin, F. S., III, Torn, M. S., & Tateno, M. (1996). Principles of ecosystem sustainability. *The American Naturalist, 148*, 1016–1037.

Clark, W. C. (2003). *Sustainability science: Challenges for the new millennium.* An address on 4 September at the official opening of the Zuckerman Institute for Connective Environmental Research and the Third Sustainability Days, 4–10 Sept 2003, University of East Anglia, Norwich. http://sustainabilityscience.org/ists/docs/clark_zicer_opening030904.pdf. Accessed April 15, 2011.

Clark, W. C., Crutzen, P. J., & Schellnhuber, H. J. (2004). Science for global sustainability. Toward a new paradigm. In H. J. Schellnhuber, P. J. Crutzen, W. C. Clark, M. Claussen, & H. Held (Eds.), *Earth system analysis for sustainability* (pp. 1–28). Cambridge: MIT Press.

Holland, A. (2005). Sustainability. In E. Craig (Ed.), *The shorter Routledge encyclopedia of philosophy* (p. 1008). London/New York: Routledge.

Hume, D. [1740](2000). *A treatise of human nature.* Oxford: Oxford University Press.

Leopold, A. [1949/53](1966). *A sand county Almanac.* Oxford: Oxford University Press.

Newton, L. H. (2003). *Ethics and sustainability. Sustainable development and the moral life.* Upper Saddle River: Prentice Hall.

Norton, B. G. (2005). *Sustainability. A philosophy of adaptive ecosystem management.* Chicago/London: The University of Chicago Press.

Ott, K., & Thapa, P. (Eds.). (2003). *Greifwald's environmental ethics.* Greifswald: Steinbecker verlag Rose.

Quaas, M., Baumgärtner, S., Becker, C., Frank, K., & Müller, B. (2007). Uncertainty and sustainability in the management of rangelands. *Ecological Economics, 62*, 251–266.

Simpson, J., & Weiner, E. (Eds.). (1989). *The Oxford english dictionary* (2nd ed.). Oxford: Oxford University Press.

United Nations (UN). (1992). *Agenda 21.* New York: United Nations.

United Nations (UN). (2000). *United Nations millennium declaration*, General Assembly resolution 55/2 of 8 September 2000. New York: United Nations.

United Nations (UN). (2002). *Report of the world summit on sustainable development.* New York: United Nations.

van den Berg, J. C. J. M., & Hofkes, M. W. (1999). Economic models of sustainable development. In J. C. J. M. van den Berg (Ed.), *Handbook of environmental and resource economics* (pp. 1108–1122). Cheltenham: Edward Elgar.

World Commission on Environment and Development (WCED). (1987). *Our common future.* Oxford: Oxford University Press.

Weber, M. [1918](1988). Der Sinn der "Wertfreiheit" der soziologischen und ökonomischen Wissenschaften. 1917. In J. Winckelmann (Ed.), *Max Weber. Gesammelte Aufsätze zur Wissenschaftslehre* (pp. 489–540). Tübingen: Mohr.

Chapter 3
The Inherent Ethical Dimension of Sustainability – Toward a Relational Ethical Perspective

By its very meaning, the modern concept of sustainability has an inherent ethical dimension. This dimension is related to the orientational aspect of the term—to its normative and evaluative meaning. Orientation means to give a direction, to guide one's actions, to distinguish between what is right and wrong, to say how one ought to act and live. Generally, orientation can be given by different means, such as power and leadership or religious faith. However, within an academic approach, we want to be able to systematically refer to the issue of orientation by reason and argumentation. The respective academic discipline that deals with orientation in this way is philosophy. Normative and evaluative issues are traditionally the subject matter of the philosophical subdiscipline *ethics*. Ethics deals with the analysis of normative and evaluative issues, with the questions of how one ought to live and how one ought to act.

In regard to sustainability, there are several normative questions that need to be analyzed. For instance: What systems, processes, or entities ought to be maintained? For whom do we have to maintain them, and for what reasons? We need to further construe the orientational dimension of sustainability and to discuss its reasons. We must not only analyze *how we can* maintain certain systems, processes, or entities, but also *why we should* maintain them. Is there any responsibility or obligation to maintain a certain system, process, or entity? How could we justify and specify such responsibilities or obligations? How could we give evidence for them?

The challenge of sustainability is to find an appropriate ethical approach to deal with these questions. We need a sustainability ethics that is able to address the specific normative and evaluative aspects of sustainability; that is able to address and analyze its orientational meaning. The development of such a sustainability ethics and its integration into an encompassing sustainability research is the main purpose of this book. The first crucial task in this regard is to identify and explain the specifics of the orientational meaning of sustainability.

For this, it is important to understand in what ways the normative connotation is grounded within the concept of sustainability itself. As long as we focus merely on the aspect of continuance incorporated in the meaning of sustainability, we cannot recognize this normative grounding, and will be rather misled. The orientational aspect of sustainability is not linked to the aspect of continuance but rather to the

C.U. Becker, *Sustainability Ethics and Sustainability Research*,
DOI 10.1007/978-94-007-2285-9_3, © Springer Science+Business Media B.V. 2012

reference to relationships. Continuance is not a value or norm in itself. Every system, process, or entity could become a matter of continuance. However, there are many systems, processes, or entities the continuance of which would not be considered positive or imperative, such as, for example, certain viruses. In regard to sustainability we need to know what systems, processes, or entities we should continue, and for what reasons. A criterion for decision cannot be found within the continuance aspect itself, but through the relational aspect of sustainability. The sustainability relations are the only fruitful and appropriate basis for construing the inherent origin and determination of the orientational meaning of the modern sustainability concept. It is only in regard to its relevance for the sustainability relations that the continuance of certain systems, processes, or entities can become a meaningful and justifiable imperative. Further ethical analysis of the orientational meaning of sustainability must, therefore, refer to the relational meaning of the term.

Some ethical aspects of the sustainability relations have already been recognized and addressed in recent discussions in different fields of applied ethics. One has, for example, prominently referred to the concept of justice and discussed issues of intergenerational justice (see, e.g., Sikora and Barry 1978; Barry 1997) as well as of intra-generational and environmental justice (see, e.g., Wenz 1988; Figueroa and Mills 2003). There have been particularly intensive discussions about the question of if and how far we have responsibilities or obligations to future generations (see, e.g., Sikora and Barry 1978; Jonas 1979; Partridge 1980), or directly to nature (see, e.g., Regan 1983; Taylor 1986). However, justice, responsibility, and obligation are particular ethical concepts, and most contributions have applied them not to all, but only to some of the sustainability relations. An encompassing sustainability ethics requires a more systematic approach to the ethical dimension of sustainability as a whole, that is, an integrated approach to all three sustainability relations and their ethical aspects and potential ethical conflicts. I suggest, therefore, not to start with rather specific questions about certain obligations, responsibilities, or justice in regard to one or two sustainability relations, but rather to address the ethical aspects of sustainability in a systematic way by basing a sustainability ethics on an encompassing analysis of *all three* sustainability relations together. This is to develop an ethical framework that allows us to simultaneously address the ethical aspects of the threefold relatedness brought up by sustainability in an *integrated way*. In this respect, the challenge of sustainability ethics is to develop an integrated type of *relational* ethics.[1]

What is the basic ethical question in regard to the sustainability relations? For an answer, one first needs to define ethics and ethical issues in general. What is ethics about? If we refer to main ethical theories that have dominated the last two centuries, we find different answers. Deontology—the ethical theory mainly referring to

[1] The term "relational ethics" is already used within current philosophy to denote a specific ethical theory based on feminism. Although I will refer to this theory later on, I do not intend to use the term "relational ethics" in this sense, but rather in the general sense of *an ethics founded in relationships and their specific characteristics.*

the tradition of Kantian philosophy—concentrates on absolute, reasonable principles for right actions. Utilitarianism discusses the question of right action based on the overall consequences of actions and resulting overall utility or happiness. Virtue ethics, in contrast, focuses on these questions: What person ought one be? What is a good character? What is a good life? For the purpose of developing a sustainability ethics, I suggest the following general definition of ethics, which encompasses all the more specific questions of the established ethical approaches: *Ethics is the systematic reasoning about the question: How ought one to live?*

Considering the specific relational meaning of the modern concept of sustainability (Fig. 2.1), the basic ethical question in regard to sustainability then is: *How ought one to live in regard to the sustainability relations*? An alternative formulation would be: How ought one to live in regard to one's embedment in the threefold relationship with contemporaries, future generations, and nature? Sustainability ethics must be able to analyze this question in a systematic way. This is a demanding and difficult philosophical task for several reasons. One main reason is that this basic ethical question comprises three very different ethical sub-issues in regard to the three sustainability relations:

(i) The moral relationship between humans and their contemporaries
(ii) The moral relationship between humans and future generations
(iii) The moral relationship between humans and nature

Sustainability ethics must be able to deal with three very different relationships that show very different characteristics and ethical aspects. Sustainability ethics must be able to simultaneously address ethical aspects of all three sustainability relations and their constitution. It must be an integrated analysis of the ethical aspects of the threefold embeddedness of human beings in the sustainability relations. Particularly, sustainability ethics should enable us to identify and analyze ethical conflicts and trade-offs among the three sustainability relations, and should provide ways to adequately integrate them.

Before proceeding with the development of an approach of sustainability ethics, I will discuss the potential and limits of traditional ethical theories and current ethical trends for approaching the sustainability relations. I argue that we cannot just use what we already have and apply existing ethical theories without any modification to the issues of sustainability, but that we rather need to develop a new specific type of sustainability ethics.

References

Barry, B. (1997). Sustainability and intergenerational justice. *Theoria, 45*, 43–65.

Figueroa, R., & Mills, C. (2003). Environmental justice. In D. Jamieson (Ed.), *A companion to environmental philosophy* (pp. 426–438). Oxford: Blackwell.

Jonas, H. (1979). *Das Prinzip Verantwortung. Versuch einer Ethik für die technologische Zivilisation*. Frankfurt: Insel.

Partridge, E. (Ed.). (1980). *Responsibilities to future generations*. Buffalo: Prometheus.

Regan, T. (1983). *The case for animal rights*. Berkeley: University of California Press.
Sikora, R. I., & Barry, B. (1978). *Obligations to future generations*. Philadelphia: Temple University Press.
Taylor, P. (1986). *Respect for nature. A theory of environmental ethics*. Princeton: Princeton University Press.
Wenz, P. (1988). *Environmental justice*. Albany: State University of New York Press.

Chapter 4
Limits and Potential of Traditional Moral Philosophy and Current Ethics – Some Arguments for the Need for a New Type of Sustainability Ethics

The philosophical tradition offers elaborate ethical theories, and some important approaches have been developed by recent ethics, too. It would be the most convenient option simply to apply one of them to the ethical questions of sustainability. However, I will give some arguments why this would be neither a fruitful nor an appropriate approach, and why we need a new type of sustainability ethics instead. This does not mean, however, that we have to completely abandon every insight of traditional moral philosophy or recent ethics. Although we cannot directly apply one of the established ethical theories, we can get some important hints from them for the project of sustainability ethics. This holds true in particular for environmental ethics, feminist philosophy, and virtue ethics.[1]

4.1 The Limits of Utilitarianism and Deontology

Utilitarianism and deontology have been the two most established ethical theories within moral philosophy over the last two centuries. Both have been developed mainly for the analysis of interactions between humans, and the focus has been on humans who live and act together in the same time. In terms of the sustainability relations this means that traditional moral philosophy has mainly focused on relation (i), that is, on relationships among contemporaries. This does not necessarily imply that it is not possible to apply established ethical theories to the other sustainability relations, but it indicates that difficulties may arise.

[1] I restrict the discussion here to exemplary considerations on some main ethical approaches and characteristics. I hold, however, that some other important ethical theories, such as Rawls' contract theory or discourse ethics, exhibit similar issues and limits in regard to the ethical dimension of sustainability. For some discussion extending to other ethical theories and characteristics, see, e.g., Palmer (1994) and Ott and Thapa (2003).

C.U. Becker, *Sustainability Ethics and Sustainability Research*,
DOI 10.1007/978-94-007-2285-9_4,

It is important to recognize that there are substantial differences between the relationship among contemporaries and the other two relationships addressed by sustainability. Both the relationship between currently living humans and future generations and the relationship between humans and nature show specific characteristics. With future generations we do not have a direct relationship (if we assume a more distant future and not just the immediately following generation). We do not know who will exist in the distant future and cannot communicate directly with future people. We do not have direct access to their interests, opinions, values, etc. In this respect, it is an asymmetric and abstract relation. We may be able to imagine some of their interests. However, it still remains a rather abstract relationship with people who do not yet exist and whom we cannot know and directly approach. The relationship is asymmetric because we can affect the future by our decisions and actions today, but not vice versa. Even if we consider a direct relation with future generations, such as the relation with our children or grandchildren, this still remains an asymmetric relation to some extent, because we cannot approach and communicate with younger children the same as we can with adults.

Even more specific and difficult is the human–nature relation. Here, we also have a distinct and asymmetric situation. We cannot approach and communicate with most entities in nature in the same ways that we approach and communicate with other humans. Moreover, we do not even know how this relation can be adequately defined in regard to its moral aspects. We have to clarify some fundamental philosophical questions to be able to define this relationship: What is nature? What do we have in common with nature? What distinguishes us from nature? How could we define a moral relationship with nature? (See, e.g., Becker and Manstetten 2004)

To what extent can established ethical theories deal with the different characteristics and specific challenges of the two sustainability relations—our relation with future generations and with nature? I first refer to some possibilities and limits of utilitarianism in this respect. It has already been recognized that utilitarianism has some difficulties in dealing with the ethics of the relationship between humans and future generations (see, e.g., Sikora and Barry 1978; Partridge 1980; Parfit 1984). The consequentialistic design of utilitarianism causes several problems: What will be relevant consequences of certain actions in the distant future, which people will be affected by them, and in what way will they be affected? What will their interests be? Can we, for example, exactly predict the long-term consequences of genetic engineering? Who will be affected by certain genetically modified organisms at what point in time in the future, and in what way? In an increasingly complex world the prediction of the long-term consequences of our actions becomes increasingly difficult. Thus, the very approach of basing ethical considerations on the consequences of actions—as is favored by utilitarianism—seems rather problematic and not fully applicable to our relationship with future generations. The features of hedonism and welfarism, which also characterize utilitarianism, cause some additional problems: In which way should future generations be included? Should, for example, their interests and happiness be discounted, as has prominently been argued within utilitarianism (see Bentham [1781]1988: iv.8)? This would mean that the interests of generations living in the farther future would count in effect for nothing in today's ethical considerations, and one might doubt that this is an adequate approach to the ethical aspects of the relationship between currently living humans and future generations.

To the human–nature relation, in contrast, utilitarianism has been applied with some "success" by extending the fundamental criterion of pain and pleasure to other nonhuman species. It has prominently been argued that there are nonhuman species that are capable of feeling pain and pleasure, and which, thus, have to be included in ethical considerations and the calculation of the greatest happiness of the greatest number (Bentham [1781]1988: xii.1.4 with footnote, Singer 1975). However, in this case, the human–nature relation is narrowed down to the relationship between humans and certain other "higher" animals, that is, those species to which we can sensibly ascribe the capability of feeling pain and pleasure (Singer 1975). This means a crucial reduction of the ethical issue of the human–nature relationship. Within both environmental ethics and sustainability ethics we are ultimately interested in the question of our moral relationship with nature in *all* its parts and features and not merely in the reduced question of our moral relationship with some higher animals. This may be an interesting and consequent extension of the utilitarian theory and may provide some moral insights. However, in this case, the ethical problem has been defined in regard to the theory. For the topic of sustainability it would be more appropriate to design the ethical approach in regard to the problem—in regard to the human–nature relation as a whole.

Deontology may avoid some difficulties in respect to the ethics of the relationship between contemporaries and future generations. Deontology does not refer to the consequences of actions, but rather to the rational motivation and justification of actions. How one ought to act is defined by reason. Reason is able to determine the fundamental ethical guidelines, and this is the basis for the relation between all rational beings (Kant [1785]1998, [1788]1998). This approach, therefore, does not necessarily distinguish between now living rational beings and rational beings living in the future. This allows, in principle, just to extend the ethical insights of deontology to the relationship with future generations. For instance, the respect of the dignity of the human being, which Kant ([1785]1998: BA 76–78) deduces in his ethical theory, holds for all human beings—contemporaries and future generations—in the same way.

However, the reason-based design entails some fundamental problems in regard to the application of deontology to the human–nature relation. Originally, Kant focused his ethical considerations on the role of practical reason for moral guidance and the implications for the relationship between human beings. Beings that are not capable of reason—that is, most parts of nature—are excluded from ethical considerations. They are not part of the moral community. Even if it would be possible to design a modified deontological approach that allows for moral guidelines for the human–nature relation, there remains a fundamental separation of humans and nature. By the very design of deontology the human–nature relation is reduced to a relation between rational beings and nonrational beings. The "oughts" in regard to this relation can merely be defined as laws and restrictions given by human reason. Other aspects of the human–nature relation beyond the mere rational approach can not be taken into account (see Part III for a more detailed discussion of this issue).[2]

In summary, a simple application of the established ethical theories of utilitarianism and deontology to all of the sustainability relations causes problems. None of them are capable of addressing the specific characteristics of both of the relations (ii) and (iii) in an adequate way. Thus, although these established ethical theories may provide some insights for some aspects and limited cases of sustainability issues, it is not feasible to use them for an encompassing ethical analysis of the threefold embeddedness of human being in the sustainability relations. Utilitarianism and deontology are not appropriate to fully discuss the ethical dimension of the threefold relatedness of human being in an integrated way.

4.2 Environmental Ethics and Sustainability Ethics

One may wonder whether some of the various fields of applied ethics that have been developed within the last 30 years may cover the ethical issues of sustainability or offer a theoretical basis for sustainability ethics. Environmental ethics would seem one potential candidate, but also, for example, development ethics. However, I will argue that particularly environmental ethics is substantially different, and should be distinguished, from sustainability ethics. The same holds for other existing fields of applied ethics.

Environmental ethics has emerged in the second half of the twentieth century and is a heterogeneous field that encompasses a variety of different ethical approaches to environmental issues. Environmental ethics is not an applicable basis for sustainability ethics. Most of environmental ethics refers primarily and in rather specific ways to the third sustainability relation—that is, the human–nature relation. Despite the diversity of environmental ethics, many contributions to the field have in common that they are based on applications of established moral philosophy (see Palmer 1994). A prominent example is the utilitarian approach of Peter Singer (1975, 1993). With this, much of environmental ethics inhibits the same theoretical problems of applying established moral philosophy to the issue of sustainability that I have discussed in the previous section. This means, in particular, that the subject matter—that is, the human–nature relation—has in many cases made fit for a specific ethical theory rather than a theory has been developed for the subject matter. In applying established ethical theory to the human–nature relation, those aspects of the

[2] It is important to note that I am using here a rather strict definition of deontology in the tradition of Kant. Some broader definitions just require that an ethical approach constitutes any kind of duty or individual dignity without referring to consequences to be a deontological approach. We then may evaluate the limits and potential of deontology to approach the issue of sustainability in a different way. For such an interpretation in regard to environmental ethics, see, e.g., Palmer (1994).

relationship are singled out that fit to the ethical theory chosen. By this, one focuses only on specific aspects of nature and of the human–nature relation that are appropriate for the ethical theory one wishes to apply. For instance, Singer (1975) singled out animals that can feel pain and pleasure, Regan (1983) vaguely singled out higher animals, and Taylor (1986) referred to all organisms. Such approaches reduce nature and the human–nature relation in certain ways. They do not deal with the question, "What is our moral relation with nature?" in an encompassing manner. They rather reduce this question to the question, "What is our moral relation with the specific part x of nature?," choosing a part x that is suitable for the application of the ethical theory chosen.

These types of approaches are based on the assumption that a moral relation can be ascribed only to individual beings. Therefore, these approaches are often labeled *individualistic* approaches. However, there are also so-called *holistic* approaches, such as deep ecology or land ethics, which, in contrast, regard nature as a whole as the subject matter of environmental ethics. The holistic approaches, however, tend to neglect the difference between humans and nature, as well as the individual and other self in nature. By this, the holistic approaches miss a full recognition of nature just as the individualistic approaches do by their neglect of nature's integrity and unity (Becker and Manstetten 2004).

Despite the fundamental dispute about the appropriate definition of the human–nature relation as a subject matter of ethical considerations, it is not always clear whether this relation is the exclusive subject matter of environmental ethics at all. Some approaches in the field of environmental ethics refer also to relation (i)—the relation with other contemporaries—insofar as this relation is mediated by nature. This holds, for instance, for the whole debate on environmental justice. Furthermore, many approaches within the field of environmental ethics refer to relation (ii)—the relation with future generations, for example, all contributions to discussing the issue of intergenerational justice in the context of environmental issues.

However, as far as I can see, there is no approach within the field of environmental ethics that systematically refers to all three sustainability relations simultaneously and in an integrated way.[3] In fact, most approaches do not give an explicit methodological consideration on their subject matter, and the field of environmental ethics as a whole shows some lack of methodological reflection in this regard. For the sake of clarity, I am using the label "environmental ethics" merely for the ethics of relation (iii)—that is, for the analysis of the question "What is our moral relation with nature?" From this, I distinguish sustainability ethics, which simultaneously deals with all three sustainability relations. In other words, the subject matter of environmental ethics is the ethical dimension of the human–nature relationship, whereas the subject matter of sustainability ethics is the ethical dimension of the threefold relatedness of human being defined by the sustainability relations. Likewise, I would distinguish sustainability ethics from development ethics. I consider development ethics as mainly

[3] Some approaches extend to two of the sustainability relations and discuss potential trade-offs between them, for instance, Rolston (1994) and Attfield (1998).

focused on the relationship among contemporaries on a global level and on specific ethical issues of this relationship. This may include environmental aspects, but not always and not in a systematic way.

The ethical questions surrounding the issue of sustainability require an ethical approach which, from the very beginning, is designed to adequately and simultaneously address all three sustainability relations and their constitution in an integrated way. Only this will allow a systematic and encompassing approach to the ethics of sustainability, and, in particular, the adequate discussion of conflicts between, and adequate integration of, these relations. Such an encompassing sustainability ethics has to start with a detailed discussion of the specific characteristics of each of the sustainability relations and the interrelations and connections between them (see Part III).

The discussion so far has demonstrated that it would be an overly simplified approach to just apply established ethics to the issue of sustainability. However, attempting to develop a new ethical theory for sustainability completely beyond the tradition of philosophical ethics would also be inappropriate. Several problems have been addressed and insights developed in the tradition of philosophy that are relevant to the project of sustainability ethics. This holds, in particular, for the tradition and recent development of virtue ethics as well as for recent contributions by feminist philosophy.

4.3 Virtue Ethics and Ethics of Care: The Ethical Relevance of Relationships

So far, I have mainly referred to utilitarianism and deontology as established ethical theories. However, there is a third ethical theory that has a much longer tradition and has become of specific importance for recent ethical discussions, and that is of particular relevance for the ethical analysis of sustainability: *virtue ethics*.[4] Virtue ethics has main historical roots in the philosophy of Aristotle and has seen a revival as an alternative to the mainstream ethics of utilitarianism and deontology in the second half of the twentieth century. In contrast to utilitarianism and deontology, virtue ethics does not focus on right or wrong actions. Rather, it focuses on the person, her character and life, and analyzes questions of what is a good character and what constitutes a good life. This focus of virtue ethics was seen as a fruitful alternative to the combating theories of deontology and utilitarianism, which have been criticized for excessively narrowing down the ethical problem to a focus on rightness of single actions, resulting in a neglect of the person, her character, life, and relationships.[5]

---

[4] Newton (2003) also bases her ethical discussion of sustainability on virtue ethics and offers some arguments why this is the most adequate approach.

[5] Anscombe (1958) is widely regarded as being the initial article for the revival of virtue ethics in the twentieth century. A further prominent contribution was made by MacIntyre (1985). An overview of the discussion can be found in Crisp and Slote (1997) and Darwall (2002).

For the development of sustainability ethics, one characteristic of virtue ethics is of particular interest, namely its explicit reference to the ethical meaning of relationships. The crucial relevance of relationships for ethics was originally recognized by Aristotle and, although he focused on specific close relationships between contemporaries, his considerations offer some important insights for the project of sustainability ethics and the ethical aspects of the sustainability relations.

Aristotle (1995, 2000) understood the human being as a rational being (a *zoon logon echein*) and political being (a *zoon politicon*). The latter characteristic means that humans are necessarily social beings who need to live together with other humans. Therefore, ethical considerations must not refer to an autonomous individual, but to the person embedded in a community, that is, in certain relationships between contemporaries. Ethics has to analyze the questions of a good life, person, and character in regard to this embedment of the person in relationships. Aristotle's analysis refers to the specifics of close relationships between contemporaries in a small state (*polis*). For Aristotle, the good life can only be achieved by acting as a virtuous person within a polis. A virtue is an acquired habitus that capacitate for good actions and good interactions with other people. Virtues cannot be recognized and acquired by an abstract theoretical insight, but rather need to be developed over time by experience of various situations and encounters. By acting in concrete situations and relationships, the initial emotions and drives of a person can be cultivated into certain virtues with the assistance of rationality (*logos*). However, what are virtues cannot be defined by theoretical rationality, but only by practical wisdom (*phronesis*) referring to the specifics of concrete situations of action. The development and definition of virtues particularly refers to the community and its specific relational structures in which the person is living and acting. This is an interesting philosophical point, because it specifies the ethical relevance of relationships: They constitute the framework in which a good life and virtues can be realized. Moreover, what is good, what is a virtue, needs to be determined by practical wisdom and by feedback and consideration of the community in which the person is living and acting (Rese 2003; Becker 2009).

The insight in the relevance of relationships for ethics has been recognized in recent ethical discussions as well. This was partly in reference to Aristotle, for example, prominently by MacIntyre (1985), who stresses the general insight of virtue ethics into the essential interconnection between virtues and relationships (MacIntyre 1985: 181–203). In addition, there are also recent approaches by feminist philosophy that emphasize the ethical meaning of relationships. Feminist philosophy has focused prominently on the ethical aspects of relationships, such as the relationship between parents and children, which has led to the rise of a specific relational ethics: the *ethics of care*. Ethics of care demonstrates that many moral issues essentially depend on the specific characteristics of relationships. Moral issues cannot just be analyzed and determined in a general and abstract way, but require an analysis of the specifics of the relationships in which they are actualized. Ethics of care takes into consideration that morality falls short if it conceives of human beings just as abstract, equal, and autonomous individuals, but must instead recognize humans as related, dependent beings embedded in many "unchosen relations" of "unequal power" (Held 2006: 46).

Both virtue ethics and ethics of care provide us some insights about the ethical relevance of relationships and how to analyze them, although both mainly focus on rather specific and close relationships between contemporaries. They provide us with the general insight that the human being cannot just be seen as an independent and autonomous individual, but that it is, to an important extent, a dependent and relational being embedded in several relationships by necessity. An adequate ethical analysis of human morality requires taking this into account. I will refer to this background when developing the relational dimension of sustainability ethics in Part III.

4.4 The Ethical Relevance of Social Structures and Institutions

There is a further, maybe even more important philosophical insight for the project of sustainability ethics provided by virtue ethics and feminist philosophy. It concerns the crucial meanings of social structures and institutions for a relational ethics.

This aspect has already been discussed by Aristotle. He recognized that human relationships need to be organized and managed and that this requires adequate social structures and institutions. Because virtues and social relationships are interconnected, the ethical issue also becomes an institutional and structural issue. The proper organization and development of the community becomes part of the ethical issue. Ethics is not just about the individual, but also about the optimal support of individual morality by adequate social institutions and structures. Aristotle concentrated on the close relationships he considered crucial for ethical issues in the context of his time. With this, he focused on two social structures he considered ethically relevant: the household (*oikos*), as a more basic organization of very elemental relationships, and the state (*polis*), as the ultimate structure of organizing the relationships between free citizens. The proper organization of both of these structures is not merely an economic and political issue, but also a crucial ethical one (Becker 2009). This is because the proper organization of these structures allows for the realization of human relationships in an ideal way, which enables the individual to be a good person and to live a good life within them. In his *Politics*, Aristotle gave a detailed analysis of the *polis* and the *oikos*. According to Aristotle, the *oikos* is the institution that integrates and guarantees the stability of three elemental relationships between humans: The relationship between husband and wife, master and slave, and father and children[6]. The *oikos* has to be organized and managed to maintain their stability and ideal realization. This is the subject matter of the *oikonomia*, the household management. The household manager—that is, the master of the house—is responsible for managing the house, which means he has to recognize the different relationships and their specifics, and foster the development of the virtues

[6] These are, of course, rather specific relationships that we would not regard as relevant today. We are, therefore, not interested in the specifics Aristotle presents in regard to these relationships, but rather in his general philosophical approach and insights.

of all members of the household according to the relationship they are in (Aristotle 1995: 1260a 14–24). More important, however, is the proper design of the state. Only a certain design, which allows the free and equal intercourse of the citizens, will lead to a mutual perfection of the virtues and the community as a whole. Thus, for Aristotle, the answer to the question of how one should live depends also on the proper organization of human relationships, and this requires the adequate development of institutions such as the *oikos* and the *polis* (Becker 2009).

In a rather different way, feminist philosophy has recognized the relevance of structural aspects for relationships and ethics as well. Here, the structural aspect is generally addressed in a more abstract way in referring to fundamental patterns of thought and action—to gender patterns, which essentially influence the concrete constitution of human relationships and their morality. Feminism identifies a theoretical structure of thought—a logic of domination—that underlies the practical historical domination of women by men. This logic is grounded on two basic assumptions: A basic dualism between men and women, which distinguishes both by a set of characteristics and states the "otherness" of women, and a basic evaluation ascribed to this dualism, which evaluates one side—men—higher than the other side—women. This pattern of thought has historically been implemented into practical life over time by a set of institutions that are designed accordingly. The very way in which societies, education, marriage, business, politics, etc., have been historically designed has determined the roles as well as the individual potentials for thought and action of women and men. It is this whole set of institutions that has stabilized the gender roles over time. The institutional framework led to an automatic adoption of gender patterns by the individuals, which were set into these given institutions and social structures. With this, the social structures govern the relationship between women and men and affect their morality (see, e.g., Beauvoir [1949]1972).

Virtue ethics and feminist philosophy both indicate, in rather different ways and on different levels of abstraction, that there is a crucial systematic connection between social relationships and social structures. We can learn from both that if we want to analyze the ethical meaning of relationships, we must also analyze the structural framework that governs and supports these relationships. In other words, a *relational* ethics must necessarily be also a *structural* ethics.

Although both virtue ethics and feminist philosophy have been recognized and introduced into the debate on environmental issues, the lessons that could be learned from them have not fully been considered so far. The small movement labeled *environmental virtue ethics* explicitly applies virtue ethics to environmental issues.[7] However, the focus of this movement is more on concrete environmental virtues and on the meaning of virtues and human excellence within the human–nature relation (see Sandler and Cafaro 2005). Environmental virtue ethics does not provide an

[7] A good overview of the recent discussion is provided by Sandler and Cafaro (2005). However, there have been important forerunners to this discussion in the period of Romanticism, e.g., Novalis (see Becker and Manstetten 2004) and Henry David Thoreau (see Cafaro 2001; Becker 2003).

encompassing analysis of the institutional and structural aspects relevant for the human–nature relation, nor does it provide a theoretical framework for referring to all sustainability relations in an integrated way. *Ecofeminism*, which is based on feminist philosophy, mainly provides an extension of this philosophy and discusses the gender pattern as crucial structure and explanatory basis for environmental issues, for example, by interpreting the human–nature relation as a further expression of the logic of domination originally found in the relationship between men and women (see, e.g., Warren 1990, 1994).

Neither environmental virtue ethics nor ecofeminism have fully recognized the fundamental insights that can be gathered from virtue ethics and feminist philosophy for the issue of sustainability, that is, the general relevance of relationships and structures for sustainability ethics.[8]

References

Anscombe, G. E. M. (1958). Modern moral philosophy. *Philosophy, 33*, 1–19.
Aristotle. (1995). *Politics* (Books I and II) (T. J. Saunders, Trans.). Oxford: Clarendon Press.
Aristotle. (2000). *Nicomachean ethics* (R. Crisp, Ed. & Trans.). New York: Cambridge University Press.
Attfield, R. (1998). Saving nature, feeding people, and ethics. *Environmental Values, 7*, 291–304.
Beauvoir, S. [1949](1972). *The second sex* (H. M. Parshley, Ed. & Trans.). Harmondsworth: Penguin.
Becker, C. (2003). *Ökonomie und Natur in der Romantik. Das Denken von Novalis, Wordsworth und Thoreau als Grundlegung der Ökologischen Ökonomik*. Marburg: Metropolis.
Becker, C. (2009). Logos und Wirtschaft bei Aristoteles. Ein dogmenhistorischer Beitrag zur Diskussion des ökonomischen Rationalitätsbegriffes. *Archives for Philosophy of Law and Social Philosophy, 95*, 523–539.
Becker, C., & Manstetten, R. (2004). Nature as a you. Novalis' philosophical thought and the modern ecological crisis. *Environmental Values, 13*, 101–118.
Bentham, J. [1781](1988). *The principles of morals and legislation*. Amherst: Prometheus.
Cafaro, P. (2001). Thoreau, Leopold, and Carson: Toward an environmental virtue ethics. *Environmental Ethics, 23*, 3–17.
Crisp, R., & Slote, M. (Eds.). (1997). *Virtue ethics*. Oxford: Oxford University Press.
Darwall, S. L. (Ed.). (2002). *Virtue ethics*. Oxford: Blackwell.
Held, V. (2006). *The ethics of care. Personal, political, and global*. Oxford: Oxford University Press.
Kant, I. [1785](1998). *Grundlegung zur Metaphysik der Sitten*. Frankfurt: Suhrkamp.
Kant, I. [1788](1998). *Kritik der praktischen Vernunft*. Frankfurt: Suhrkamp.
MacIntyre, A. (1985). *After virtue*. London: Duckworth.
Newton, L. H. (2003). *Ethics and sustainability. Sustainable development and the moral life*. Upper Saddle River: Prentice Hall.

[8] See, however, Sherwin (2008) for a more demanding and comprehensive programmatic call for a new kind of bioethics based on feminist relational theory. Sherwin particularly identifies the crucial ethical meaning of social institutions and organizations for issues of bioscience and argues for a *public ethics* that shall explicitly refer to the ethics of institutions and organizations, and its relation to individual morality.

Ott, K., & Thapa, P. (Eds.). (2003). *Greifwald's environmental ethics*. Greifswald: Steinbecker verlag Rose.

Palmer, C. (1994). A bibliographical essay on environmental ethics. *Studies in Christian Ethics, 7*, 68–97.

Parfit, D. (1984). *Reasons and persons*. Oxford: Oxford University Press.

Partridge, E. (Ed.). (1980). *Responsibilities to future generations*. Buffalo: Prometheus.

Regan, T. (1983). *The case for animal rights*. Berkeley: University of California Press.

Rese, F. (2003). *Praxis und Logos bei Aristoteles. Handlung, Vernunft und Rede in Nikomachischer Ethik, Rhetorik und Politik*. Tübingen: Mohr.

Rolston, H., III. (1994). Feeding people versus saving nature? In W. Aiken & H. LaFollette (Eds.), *World hunger and morality*. Englewood Cliffs: Prentice-Hall.

Sandler, R., & Cafaro, P. (Eds.). (2005). *Environmental virtue ethics*. Lanham: Rowman & Littlefield.

Sherwin, S. (2008). Whither bioethics? How feminism can help reorient bioethics. *International Journal of Feminist Approaches to Bioethics, 1*, 7–27.

Sikora, R. I., & Barry, B. (1978). *Obligations to future generations*. Philadelphia: Temple University Press.

Singer, P. (1975). *Animal liberation*. New York: Random House.

Singer, P. (1993). *Practical ethics* (2nd ed.). Cambridge: Cambridge University Press.

Taylor, P. (1986). *Respect for nature. A theory of environmental ethics*. Princeton: Princeton University Press.

Warren, K. (1990). The power and the promise of ecological feminism. *Environmental Ethics, 12*, 125–146.

Warren, K. (Ed.). (1994). *Ecological feminism*. New York: Routledge.

Chapter 5
The Challenges of Sustainability Ethics

From the discussion so far follows that the project of sustainability ethics combines a relational and a structural ethical challenge. However, both the relevant relationships and structures are much more difficult and complex than in the case of close personal relations, such as in the *oikos* or the *polis*, or between parents and children. In the field of sustainability ethics, we are not interested so much in the specific human relationships and related institutions on which Aristotle, other virtue ethicists, and feminist ethicists have focused. We are rather confronted with three fundamental sustainability relations of the human being, each of which possess very different and difficult characteristics. In addition, we have to take into account that today the sustainability relations are governed and influenced by a very complex set of interwoven structures. The sustainability relations are not just set up by individual behavior, attitudes, and morality, but are also governed by given structures and mechanisms, such as social and global institutions, systems, and patterns of thought and action, in which the individual is already located. The individual cannot completely independently or in isolation develop these relations, but rather understands them and acts in regard to them through a complex web of given systems and mechanisms.

For instance, how to relate oneself to nature is not an entirely autonomous individual decision, but the individual is already related to nature in many ways through given patterns of thought actions—for example, through established scientific perspectives and existing technologies. We drive cars, wear clothes, and eat food. By these everyday actions we are related to nature in many ways, without knowing in detail how. We participate in (often global) complex structures of production, distribution, and disposal that mediate the relationship between us and nature. Moreover, the attitudes and values—the very cultural identity of a society—govern the relationship its individuals have with nature. The relation to future generations is also governed by existing structures and mechanisms. For instance, the individual cannot decide in isolation what knowledge to pass down to future generations. This depends, to a large extent, on the very mechanisms and institutions society has developed for the creation, evaluation, conservation, and transfer of knowledge. In modern western societies relevant institutions in this respect are, for example, science, universities, schools, books, electronic devices for storage and communication, etc.

C.U. Becker, *Sustainability Ethics and Sustainability Research*,
DOI 10.1007/978-94-007-2285-9_5, © Springer Science+Business Media B.V. 2012

At this stage of the discussion it becomes obvious that the challenge of sustainability ethics is much more difficult and demanding than the simple application of an established ethical theory. An encompassing sustainability ethics has the task of analyzing the sustainability relations and their individual ethical challenges on the one hand, and the relevant structures and mechanisms that impact and govern them on the other hand. Sustainability ethics is concerned not only with questions of how we should understand ourselves as threefold embedded beings within the sustainability relations and how we should act and live in regard to these relations, but also focuses on questions of what structures and institutions govern these relations and in what ways, and how we should design them to allow for an ideal organization and realization of the sustainability relations. We must develop an ethical approach that fully takes into account the specifics of all three sustainability relations and the relevant structures and mechanisms that govern these relations. None of the existing ethical approaches are able to provide such an encompassing approach. We need a new sustainability ethics that is able to do so. For this, we have to start from the very beginning by analyzing the sustainability relations and the crucial structures that govern them in detail.

However, in contrast to the sustainability relations, it is not quite clear yet what the crucial structures and mechanisms are. We have already identified the sustainability relations, but the clear identification of the structures and mechanisms relevant for the issue of sustainability is still lacking. For this reason, I will proceed by focusing first on the crucial and complex issue of the structures relevant for sustainability ethics and identify, define, and analyze them in detail (Part II). Here, the challenge of the structural dimension and its characteristics will become evident and turn out to be another important reason why traditional ethics would not be an adequate means to address the ethical aspects of sustainability. The clarification of the structural aspect will provide further requirements for sustainability ethics and will allow us to proceed with its development by construing both the individual and structural dimension of sustainability ethics in Part III. After the discussion of sustainability ethics in Part III, I will proceed with discussing the project of sustainability research as a whole in Part IV.

Part II
Meta-structures and Sustainability

Sustainability is about the integrated development of our relationships with other contemporaries, future generations, and nature—i.e., about the integrated development of these three *sustainability relations*. As discussed in Part I, the ethical challenge of sustainability is a *relational* ethical challenge of how to act and live within these relationships as a threefold related being. I also have argued that this ethical challenge comprises an *individual* ethical challenge of how each individual ought to live and act in the context of the sustainability relations, and a *systemic* ethical challenge of how to properly design societal and global structures that substantially affect the sustainability relations.

Before we can proceed with a more detailed ethical discussion of these challenges, we first need to get a more precise understanding of the structural impact on the sustainability relations, i.e., we have to clarify the following questions: What structures are relevant in the context of sustainability? What characteristics do these structures have? How do they affect the sustainability relations? By what approach could we analyze these structures and their impact?

Part II deals with these questions. I refer to some previous attempts to address the structural level and develop my own framework of analysis of the structural level of sustainability. With this, Part II provides a preparation to the ethical analysis in Part III, and the methodological analyses of sustainability research in Part IV.

Chapter 6
Sustainability, Institutions, and Patterns of Thought and Action

As discussed in Chap. 4, virtue ethics and feminist philosophy already provide some evidence that sustainability cannot be merely an individual ethical challenge in the context of the sustainability relations. Rather, sustainability is also a complex structural issue, and requires an ethical approach that is capable of addressing this structural level and its ethical relevance. For this, an encompassing sustainability ethics must be able to identify and analyze the structures that have a crucial impact on the sustainability relations.

To some extent the relevance of certain structures for environmental and sustainability issues has already been recognized by academic and philosophical approaches as well as in political and public discussions. However, existing academic and political discussions of the structural aspects have some shortcomings, which impede rather than support an encompassing analysis of the issue of sustainability and its ethical aspects. Most approaches are not coherent, and the identification of relevant structures often is coincidental and arbitrary.

A crucial shortcoming on the political level is the insufficient distinction between the sustainability relations and the structures that affect them. From the very beginning of the modern discussion on sustainability, there has been a particular focus on the role of the economy for sustainability (see, e.g., WCED 1987: Ch. 3). Although the economy certainly is an important structure to be considered, the strong focus on the economy ignores other structures that are also crucial for sustainability. Moreover, this focus led to a misleading description of the whole issue of sustainability as being about how to create harmony among the social, economic, and environmental sphere.[1] This, however, confuses the sustainability relations with the structures that influence them. The social and environmental spheres are fundamental

[1] See, e.g., the Johannesburg Declaration (UN 2002: §8): "Ten years ago, at the United Nations Conference on Environment and Development, held in Rio de Janeiro, we agreed that the protection of the environment and social and economic development are fundamental to sustainable development."

C.U. Becker, *Sustainability Ethics and Sustainability Research*,
DOI 10.1007/978-94-007-2285-9_6, © Springer Science+Business Media B.V. 2012

spheres of human existence, which I have defined as part of the sustainability relations. The economy is a specific social and global structure of human action that affects the sustainability relations. For clarity of analysis it is important that we distinguish between *sustainability relations,* by which human existence is fundamentally defined, and *structures*, which affect these relations. Unfortunately, the so-called triple bottom line or three pillars of sustainability—economy, society, and ecology—, have gained much prominence within public and academic discussion up to now (Kates et al. 2005). This impedes the debate and hinders a more clear and distinct analysis of the issues of sustainability (see also Ott and Thapa 2003: 59).

A second shortcoming within the field of political and public debates on environmental issues and sustainability is the reduced focus on particular mechanisms. Several patterns of action or institutions have been addressed as main causes of environmental and sustainability problems. A broad range of arguments refers to more general structures and mechanisms, such as globalization, markets, and consumerism, or to more concrete institutions, such as the World Trade Organization (WTO), the World Bank, and big international corporations, as the main obstacles to sustainability.[2] One may wonder, however, if it is indeed one single mechanism that is crucial for environmental or sustainability issues—if, for instance, the World Bank or some large corporations are the main obstacle to sustainability, and whether, e.g., a close-down or redesign of the World Bank would substantially solve any problems. Such perspectives provide a too simplified analysis of and solution to sustainability issues. At the very least, one would have to analyze the very conception and structures of the economic system of which institutions such as the World Bank are an expression. However, the general critique of globalization or markets, which occurs in public debates as well, remains too vague and cannot provide further insights into such interconnections.

On the academic level, by contrast, there have been some more substantial analyses of fundamental structures underlying environmental and sustainability issues and affecting the sustainability relations, particularly the human–nature relation. In the field of environmental ethics, this has indeed been a rather prominent issue, and several approaches refer to certain crucial theoretical or practical patterns. *Deep ecology* has identified a fundamental pattern of thought—the subject–object division in modern philosophy—as being crucial for the modern relationship between humans and nature. According to the analysis of deep ecologists, it is this fundamental pattern of thought that sets humans in a problematic relation with nature, causes a fundamental separation and alienation of humans and nature in modern times, and is the deeper cause of a fundamental modern ecological crisis. To overcome

[2] The general critique on globalization and the global economy becomes evident in, e.g., the increasing public protests during the G7 summits and the rise of nongovernmental organizations such as Attac. There has also been a lot of criticism on the World Bank (see, e.g., Rich 1995). A rather popular critique of the institution of the corporation has, for instance, been provided by the movie *The Corporation,* by Mark Achbar, Jennifer Abbott, and Joel Bakan.

environmental problems thus requires overcoming the subject–object division incorporated in modern thought and worldview (Naess 1973, 1989; Devall and Sessions 1985; Devall 2001). *Social ecology* has referred to fundamental social structures as crucial in regard to the human–nature relationship. It identifies historically developed patterns of social hierarchy and domination as crucial in regard to environmental issues, as they lead to a domination and degradation of nature. Thus, social ecology regards the analysis and the overcoming of these social patterns as the most important tasks with respect to sustainability (Bookchin 1982, 1990). *Ecofeminism* refers to specific social patterns of thought and action as crucial for the human–nature relationship. It focuses on the gender pattern—i.e., the historical socially generated and established domination of women by men—as the most crucial social pattern and regards its analysis and overcoming as the key to understand and overcome other patterns of domination, such as the domination of nature by humans (Warren 1994).

Although these academic approaches provide some important insights to the structural level, they have shortcomings in regard to sustainability. First, they tend to reduce the explanation of environmental and sustainability problems to one single cause and to reduce the relevant structural level to one single structure or mechanism—one fundamental pattern of thought or action. Second, the patterns addressed are very general and abstract. There seems to be a gap between these general patterns and the very concrete institutions, such as markets or the World Bank, that have often been identified as relevant in public debates. The crucial question is if and how general patterns such as the subject–object division or the gender pattern realize themselves in concrete institutions such as the market economy. How do abstract patterns of thought and concrete institutions work together and form stable social structures that affect our lives and actions?

In summary, all above mentioned discussions refer to important theoretical and practical structures and societal and institutional aspects, which are of some relevance in the analysis of environmental and sustainability issues. However, each of them tends to reduce the issue to *one single* structural aspect. One single theoretical or practical aspect is typically identified as the crucial factor for environmental problems or sustainable development, and often the crucial factor remains either too specific or too general. By contrast, I argue in the following that it is a complex interplay of theoretical and practical patterns, combined and institutionalized in several distinct structures, that we must analyze and discuss within an encompassing approach to sustainability. We have to take into account that there is a complex interrelationship among theoretical assumptions, patterns of action, and institutions. Certain institutions are based on specific theoretical ideas and concepts. Certain social patterns of action are influenced by underlying patterns of thought. Both, theoretical assumptions and social patterns of action are actualized and stabilized by institutional settings. Furthermore, institutional settings can develop their own inherent dynamics, which further influence action and thought. In other words, it is neither an isolated fundamental pattern of thought nor a specific concrete institution that affects our individual and social thoughts, actions, and relationships, but rather complex conglomerates of theoretical and practical patterns working together in institutionalized frameworks. Such clusters are, for instance, the economy, science, and technology. The elements, inner structure, and dynamics of such clusters will be analyzed in the following chapter.

References

Bookchin, M. (1982). *The ecology of freedom*. Palo Alto: Cheshire.

Bookchin, M. (1990). *The philosophy of social ecology: Essays on dialectical naturalism*. Montreal: Black Rose.

Devall, B. (2001). The deep, long-range ecology movement: 1960–2000 – A review. *Ethics & the Environment, 6*, 18–41.

Devall, B., & Sessions, G. (1985). *Deep ecology: Living as if nature mattered*. Salt Lake City: Peregrine Smith.

Kates, R. W., Parris, T. M., & Leiserowitz, A. A. (2005). What is sustainable development? Goals, indicators, values, and practice. *Environment: Science and Policy for Sustainable Development, 47*(3), 8–21.

Naess, A. (1973). The shallow and the deep, long-range ecology movement. *Inquiry, 16*, 95–100.

Naess, A. (1989). *Ecology, community, and lifestyle. Outline of an ecosophy*. Cambridge: Cambridge University Press.

Ott, K., & Thapa, P. (Eds.). (2003). *Greifwald's envionmental ethics*. Greifswald: Steinbecker verlag Rose.

Rich, B. (1995). *Mortgaging the earth: The world bank, environmental impoverishment, and the crisis of development*. Boston: Beacon.

United Nations (UN). (2002). Johannesburg declaration on sustainable development. In United Nations, *Report of the world summit on sustainable development* (pp. 1–5). New York: United Nations.

Warren, K. (Ed.). (1994). *Ecological feminism*. New York: Routledge.

World Commission on Environment and Development (WCED). (1987). *Our common future*. Oxford: Oxford University Press.

Chapter 7
Meta-structures

As discussed in the previous chapter, a variety of different structures have been said to be crucial for the issue of sustainability and the sustainability relations: very fundamental patterns, such as the subject–object division or the gender pattern, and rather concrete institutions, such as markets, banks, or corporations. However, to get further insight into the way in which individual action and thought as well as the sustainability relations are influenced by all these structural aspects, we need to understand their interconnection and interplay. We cannot focus on one single pattern or institution, but rather have to consider the combination of fundamental patterns of thought and action, and concrete institutions. In what ways are fundamental patterns of thought realized and institutionalized so that they become a stable and continuing influence on actions, thoughts, and relationships? On the other hand, what fundamental patterns of thought are expressed by certain institutions or mechanisms, such as markets or globalization?

The challenge with respect to the structural dimension of sustainability is to develop a conceptual framework that captures all these systemic mechanisms. We need an approach that allows us to describe the structural impact on the sustainability relations in all aspects, to become able to provide a detailed ethical analysis of that impact. For this, I introduce the concept of *meta-structures* to capture all aspects of the structural level in one systematic framework. I define a *meta-structure* as a historically evolved structure composed of four elements—(1) basic assumptions, (2) basic evaluations, (3) driving forces, and (4) institutionalizations—that substantially affect societal and individual thoughts, actions, and relationships.

The first three elements of the meta-structure represent fundamental patterns of thought and action. *Basic assumptions* are fundamental categories of thought by which we structure and understand the world—for instance, basic concepts such as "nature" or "human being." This includes fundamental conceptual distinctions such as the subject–object division. *Basic evaluations* represent fundamental values or norms, which are often related to the basic assumptions. For instance, if there is a dualism as a basic assumption, such as that of women and men, or humans and nature, there can be an evaluation related to both parts of the dualism. One part may be evaluated as being of less value than the other, inferior to the other, to be dominated by the other, etc.

C.U. Becker, *Sustainability Ethics and Sustainability Research*,
DOI 10.1007/978-94-007-2285-9_7,

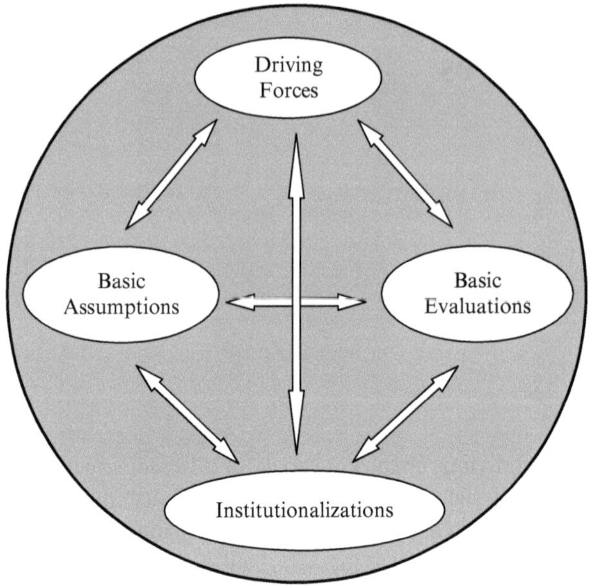

Fig. 7.1 The meta-structure and its elements

From these two elements I distinguish a third element of the meta-structure: *driving forces*, which are fundamental motivations or mechanisms underlying a certain meta-structure and causing its dynamics. This element is crucial for understanding the forces behind the development of a given meta-structure. Driving forces are often basic motivations or incentives, such as the striving for having more, the striving for power, or the striving for recognition. Motivations can be the result of a basic evaluation, but may also be grounded in irrational or emotional origins. However, other mechanisms can also be crucial to the dynamics of a meta-structure, such as incentives resulting from the very organization of the structure, e.g., the incentives given within disciplinary scientific research, which substantially drive the development of science.

Fourth, I distinguish the element of *institutionalization* of the meta-structure, which encompasses the expression, realization, and stabilization of the first three elements by concrete institutions and organizations. Institutions comprise all rules and arrangements that aim to govern human interactions and relations within a society. Organizations are functional entities that implement certain institutions. By the definition of the four elements, I particularly aim to distinguish institutions, which can be related to specific aims and areas of societal actions and relations, from basic assumptions and evaluations, which cannot be directly related to a specific aim or a specific field of action. The latter are more fundamental elements of the meta-structure that often have their origin in a fundamental given worldview.

The four elements are closely related. They build a cluster of interrelated theoretical assumptions, practical attitudes, institutions, and organizations. This cluster can be identified as a distinct entity in its own (see Fig. 7.1). This entity, the meta-structure, has its own inner consistency, stability and dynamics. The analysis of its four elements and their complex interrelation allows an encompassing understanding of the meta-structure, its dynamics, and its crucial influence on the individual, her thought, action, and relationships. Furthermore, the concept of the meta-structure allows one to distinguish different meta-structures and to analyze the interrelations among them.

I will demonstrate the fruitfulness of the concept of the meta-structure by discussing the three meta-structures I take to be crucial for the sustainability issue: *science*, *technology*, and the *economy*. This does not mean that these are the only meta-structures relevant for the issue of sustainability. However, I focus on science in particular to demonstrate the ways in which it is a complex structure of crucial influence on the sustainability relations and, with this, a matter of sustainability ethics itself. The analysis of science as a meta-structure will be important for discussing the role of science within the analysis and solution of sustainability issues in Part IV. Science is the dominant academic approach to environmental and sustainability issues, and the recognition and analysis of science as a meta-structure will shed new light on the question of how to properly design and organize sustainability research.

I additionally analyze the crucial global structures of technology and the economy to demonstrate that the concept of meta-structures (a) can be applied to other structures that affect and govern the sustainability relations and (b) allows for their identification and detailed analysis, as well. The economy and technology are crucial structures that heavily impact the sustainability relations, and their analysis will allow for a detailed discussion of the characteristics of that impact. Moreover, the analysis of all three structures will provide two further important insights. First, there are important parallels in their characteristics and impact on the sustainability relations, and second, there are complex interdependencies among science, technology, and the economy. This means that science and any other meta-structure can not fully be understood if analyzed as an isolated entity. We rather have to take into account that there is a complex web of interrelated meta-structures. Thus, the analysis of three meta-structures reveals the complexity of the structural dimension of sustainability and sustainability ethics, which has further implications for sustainability ethics and sustainability research.

7.1 Science as a Meta-structure

Science is an endeavor to generate a specific type of knowledge. Facing the challenge of sustainability, there is a common reaction to ask for more scientific research for finding answers and solutions. The latest report of the United Nations Environment Program, *GEO-4* (UNEP 2007), is an impressive example of scientific results already existing. It encompasses about 600 pages. However, the sustainability issue cannot be solved by scientific research alone. A report of more pages and of the same character will not, by itself, lead to a more sustainable development.

One reason is that there are crucial ethical and political aspects in addition to the scientific sphere, and these have to be addressed. Another reason is that science is a meta-structure in itself, with its own inherent structure and dynamics. As such, it sets us in certain patterns of thought and in a specific relationship to ourselves, other humans, and nature. In other words, science is not a neutral instrument of analysis, but rather a meta-structure that by its very design has a crucial impact on the sustainability relations by itself. However, this is not fully reflected by traditional science, and as a result science's ability to contribute to the analysis and solution of the sustainability issue is currently constricted. In this section, I demonstrate the meta-structural character of modern science and analyze this meta-structure with its elements and dynamics in detail. This will enable us to properly discuss the meaning of science in regard to the sustainability issue and to finally draw conclusion for the design of a new type of sustainability research.

There are some crucial basic assumptions and basic evaluations behind the whole project of modern science. A first fundamental basic assumption concerns the very process of generating scientific knowledge. The underlying assumption of modern science is that by empirical observation and by reason, the human mind is able to recognize nature and its laws. Humans, as recognizing rational subjects, actively approach nature as the passive object of recognition. By this, humans detect the order and laws by which nature is governed. This very understanding of scientific recognition and knowledge generation is a specific expression of the subject–object division, which is said to be fundamental to modern thought.[1] Modern science is based on the modern worldview. Science, therefore, by its very design constitutes a certain relationship between humans and nature. It incorporates a certain view of nature, of the human being, and of their relationship. If we think in scientific categories or do scientific research, we are already set in a certain relationship toward nature (Becker and Manstetten 2004).

A second important basic assumption is a fundamental *epistemological* assumption underlying modern science: the assumption that scientific research is an ongoing process of reduction of ignorance. It is generally assumed that scientific research will more and more reduce human ignorance and provide us with more and more insights into the world. The implicit understanding behind this assumption is that there is a limited given amount of ignorance that can and will be transformed into knowledge over time by scientific research (Fig. 7.2a, b). Thanks to science, less and less ignorance allegedly remains in the world.

We can, however, understand ignorance and the whole process of generating scientific knowledge in a quite different way. Science generates knowledge and raises new questions. Typically, the solution of one research problem raises several new problems and questions. Ignorance, therefore, is always on the frontier of

[1] That the dualism of subject and object is fundamental to modern (western) thought since Bacon and Descartes and causes several problems has often been stated and analyzed. A historically prominent contribution to this issue was provided by the Romantics and German Idealism. Later prominent discussion can be found, for instance, in Heidegger (see, e.g., 1950). Movements such as deep ecology also refer to this issue.

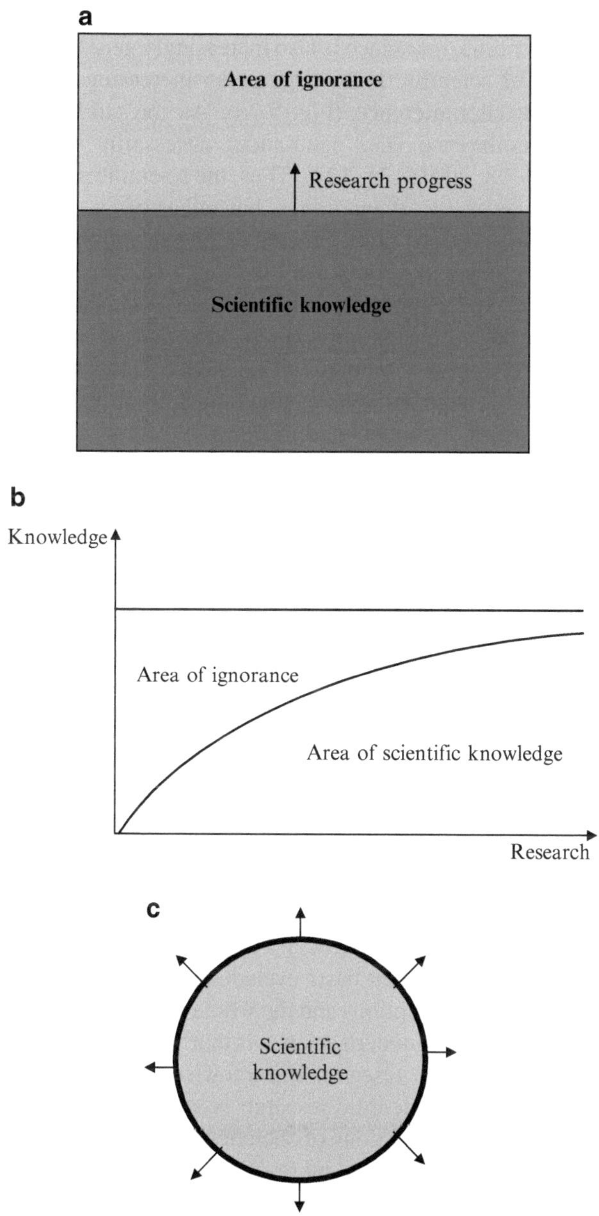

Fig. 7.2 (**a**) The traditional view of science and scientific knowledge generation. (**b**) The traditional view of science: dynamic perspective. (**c**) An alternative understanding of the dynamics of scientific knowledge

scientific knowledge, on the very edge of research. As scientific research and scientific knowledge grow over time, ignorance is also increasingly generated by this process. The picture is that of scientific knowledge as an increasing circle or ball, with ignorance being the circumference (Fig. 7.2c). As the circle (i.e., knowledge) increases, the circumference (i.e., ignorance) necessarily increases, too (see Mittelstraß 1998: 74–78, 2001: 125–127). Thus, the scientific progress is not a process of an ongoing *reduction* of ignorance, but rather of an ongoing *increase* of ignorance. This alternative view of the process of the generation of scientific knowledge is consistent with factual development in many research areas in recent times (see, e.g., Solla Price 1963; Stuhlhofer 1983).

Two remarks on these epistemological considerations: (1) I have argued that there is no "objectively" given "amount" of ignorance in the world, but that ignorance is rather created by our thought, by the human mind itself. Knowledge and ignorance cannot be sensibly understood independently from the human mind and thought. For this reason, it would, for instance, make no sense to say that the people in the Middle Ages were ignorant about quarks. The ignorance about quarks rather emerged within the development of modern physics at a certain point in time. (2) There are additional forms of ignorance created by the increase of scientific knowledge. First, the fast speed of the increase results in a loss of some amount of "older" knowledge. We are unable to keep all the knowledge that has been created over time. People and mechanisms are limited in their ability to retain all knowledge and transmit it to scientific communities and future generations. Therefore, an "inner hole" is also created in the circle or ball. This inner hole represents the loss of knowledge that did exist once and got lost over time. Second, the increasing specialization within modern scientific research increasingly leads to a situation in which merely a few specialists have knowledge of specific fields of research. This too generates an increasing ignorance within science, as more and more scientists are ignorant about many areas of their own discipline, and even more about other disciplines. I will call this the *fragmentation* of scientific knowledge.

So far, I have merely discussed crucial basic assumptions underlying modern science. However, some important basic evaluations are deeply connected to (and even embedded in) these assumptions and the whole project of modern science. The main basic evaluation behind modern science is that it is generally considered to be a positive endeavor: scientific research and knowledge are good; more scientific knowledge and, thus, more scientific research is always better then less; science helps to improve the living conditions of societies and of humankind. Science will be able to reduce ignorance more and more. With this, science will more and more enable us to control and manage the world (nature) and to better human conditions, i.e. to overcome diseases, to improve medical care, to improve agriculture and food supply, etc. This is the *Baconian ideal of science*—the *underlying implicit social contract of modern science* from the seventeenth century up to now.[2] Of course, if

[2] This basic evaluation goes back to Francis Bacon (1561–1626), particularly to his *Novum Organon* ([1620]1863). See also Sect. 7.2.

we refer to the alternative epistemological interpretation of science given above (Fig. 7.2c), we may come to a different evaluation of modern science in regard to its impact on the sustainability relations and, by this, we may come to the conclusion that we need a *redesign* of the meta-structure science or a *new* social contract for science in regard to sustainability (see Part IV).

The main driving force of modern science results from its underlying basic assumptions and evaluations. The vision that science will reduce ignorance about the world and allow for its control, and that this will improve human conditions and better human lives, has been a major incentive for scientific research up to now. However, there are further important driving forces of science. There are internal forces, resulting from the institutionalization of modern science, and there are external forces, such as economic incentives or technological influence, that also heavily affect the dynamics of the scientific meta-structure (see Sect. 7.4). The specifics of the internal forces will become evident upon closer examination of the fourth element of the meta-structure of science: the institutionalization of science.

Science is not an abstract process, but is actualized and organized in a certain way and by certain institutions and organizations. We have universities, research institutes, laboratories, national and supranational organizations financing research, etc. Modern science is organized by a set of disciplines, each of which refers to a specific set of theories and methods, and has its own standards, processes of evaluation, societies, peer-refereed journals, and so on. Specific institutional settings and their organization influence scientific research and its dynamics. Current science is not just driven by the striving for knowledge and insight, but to a large extent by the institutional settings of research and the resulting inner incentives of different scientific disciplines. For instance, methods and techniques in laboratories are often applied to as many subjects as possible—here the method is driving further research. Or, specialization is driven by the requirement to establish oneself as a distinguished specialist with one's own expertise, which often results in the making up of a specific, distinguished area of research. On the other hand, peer review is a crucial institution strongly influencing the direction of future research that often maintains established paradigms and impedes unorthodox thought and innovation. These are just a few examples of the impact the very way we have organized and institutionalized science has on the dynamics of science.

Science is a meta-structure that is set up by certain basic assumptions and evaluations, driving forces, and institutionalizations. The specific mix of elements particularly leads to the ongoing *growth* of this meta-structure. This increase is, however, not just a quantitative one resulting in more knowledge, but also has certain *qualitative* aspects. It is not just that we have more and more knowledge, but we also have more and more specialization and fragmentation within science. Therefore, we have indeed an ongoing growth of science *and* of its *complexity*. The term "complexity" is used here in a basic meaning. I define the complexity of a system by the number of its elements and the degree of their difference, and the number and degree of difference of the (internal and external) interrelations of the elements. This means that the complexity of a system increases the more elements and interrelations it has, and the more distinct both become. Increase of complexity makes it more difficult to relate the elements of a system to each other in a controlled way. At a certain point complexity even may endanger the integrity and stability of a system.

From the analysis of the meta-structure science and its elements, it has already become evident that science has an important impact on the sustainability relations. Science sets us into a specific relation with nature, other humans, and future generations. Therefore, it is crucial that we question whether this impact is in accordance with sustainability from an ethical perspective: Is the current design of science appropriate to the sustainability relations? Which elements of the meta-structure are problematic, and which are not? Do we need a redesign? What elements should be changed, and in what ways? This I will analyze in detail in Part III. Because science is an important part of the academic approach to sustainability itself, there is also the question of whether established science is an appropriate tool for the analysis for sustainability issues, or if we need a new type of sustainability research. This will be discussed in Part IV. However, before further discussing the implications regarding the meta-structure science, I will introduce and discuss two other crucial meta-structures: technology and economy.

7.2 Technology as a Meta-structure

A simple and common definition of technology is the instrumental definition: Technology is an instrument—a means—for certain ends. Technology is the extension of our ability and potential to act and to realize certain aims. However, as for science, I argue that technology is not just an instrument, but rather is also a meta-structure. As a meta-structure, technology has inherent, often hidden, basic assumptions and evaluations, its own driving forces, and a complex institutionalization. By its design and dynamic this meta-structure also deeply affects individual and social life, action, and thought, and particularly the sustainability relations.

It is again Bacon who originally provided the most prominent and dominant view of modern technology. Bacon strongly related science and technology to each other, and gave some fundamental basic assumptions and evaluation for both of them, which not only influenced the modern view of science, but also the modern understanding of technology:

> I may hand over to men their fortunes, now their understanding is emancipated and come as it were of age; whence there cannot but follow an improvement in man's estate and an enlargement of his power over nature. For man, by the fall fell at the same time from his state of innocency and from his dominion over creation. Both of these losses however can even in this life be in some part repaired; the former by religion and faith, the latter by arts and sciences. For creation was not by the curse made altogether and forever a rebel, but in virtue of that charter "In the sweat of thy face shall thou eat bread," it is now by various labors (not certainly by disputations or idle magical ceremonies, but by various labors) at length and in some measure subdued to the supplying of man with bread, that is, to the uses of human life. (Bacon [1620]1863: Aphorisms, 2, L II)

The crucial basic assumption here is that technology and science are strongly related to each other. Science provides the theoretical knowledge about nature; technology applies this knowledge and provides the practical means of action. Together, this allows for the management and control of nature. Again, this is a specific expression of the modern worldview that sharply divides subject from object. Nature is seen as a mechanical system that can be controlled and managed by humans through a combination of scientific and theoretical insights into its mechanisms and laws, *and* the respective means for action provided by technology.

The basic evaluation behind this view of technology is that it is *good* to use technology in this way—that we should understand and apply it that way. Nature is evaluated negatively. It causes pain and evil for humans. This includes, e.g., suffering from diseases, hunger, and natural disasters. However, nature is not entirely "rebellious." Nature incorporates laws that allow humans to understand, dominate, and improve it. Technology should be based on scientific knowledge and provide the means for controlling nature. The more knowledge we have, the more technology we can develop, and the more we can control nature. Science and technology together should provide the potential to control and manage nature, and, with this, improve the living conditions of humans on Earth.

There is an even more fundamental evaluative assumption behind this view, which is grounded in a certain religious worldview—in a specific theological interpretation of the biblical tale about paradise. For Bacon, the ultimate end of science and technology is to help overcome the fall from paradise and its consequences, i.e., human suffering on Earth. By using science and technology, humans shall control nature and, by this, reestablish a kind of paradise on Earth. Thus, science and technology make some form of salvation possible already in this life on Earth. The modern view that science and technology are means for improving human conditions is indeed a kind of promise of salvation and has its origins in the specific religious worldview of Bacon.

According to the Baconian view, technology is essentially driven by science, and ultimately by the striving to better human life on Earth through control and management of nature. However, although the Baconian view of technology has been very prominent and still is implicitly underlying much of the modern understanding of technology, there are several other important aspects of this meta-structure, particularly concerning its driving forces. It is not adequate to assume that technology is merely driven by science and scientific knowledge. Technology is more than just an application of scientific knowledge; it also has independent inner structure and dynamics.[3] First, it is not just theoretical knowledge that is relevant in the sphere of technology, but also know-how, which rather is a practical, personal type of knowledge. This fact has already been stressed by Aristotle, who distinguished *techne* as its own type of knowledge or rationality strictly separated from other types of knowledge or rationality (Aristotle 2000: 1140a). Bacon, who heavily argues against the Aristotelean view of science and technology, did not recognize these inner

[3] That technology shows a structure and dynamics that is independent from science has been discussed widely in the field of philosophy of technology, for instance by Ellul (1980), who argued that there is an "autonomy" of technology (pp. 125–150), and by Heilbroner (1967).

specifics of technology, which nevertheless play a crucial role in the structure and dynamics of the modern meta-structure of technology. Second, it is important to recognize that the sources of technological invention are not just scientific knowledge and discoveries. Rather, many modern technical inventions have been the result of creativity and imagination. Thus, the sphere of technology can be regarded as an expression of the human being as a creative being rather than a rational being. At its core, the important contribution of Heidegger (1962, 1977) to the philosophy of technology deals with this important aspect of technology.

To a large extent, the inner structure and dynamics of modern technology are the result of the institutionalization of modern technology. Modern technology is a complex system of interrelated and interdependent parts. The very existence of a specific single technology often leads to a broad development of further technologies. This holds in particular for basic technologies, so-called *key technologies*. For instance, the invention of the microprocessor led to many developments in all areas of technology. Another crucial aspect for the inner dynamic of the meta-structure technology is its *path-dependency*. Certain technological developments are based on and driven by previous technological developments. For instance, the whole technological progress of combustion engines for automobiles over the last 100 years is mainly driven by perfection and further development of a very specific type of engine—the Otto engine. Thus, technology has the potential to become a force in itself. The very existence of the technological system leads to improvements, further applications, and technological developments. Technology is not just a means for certain ends, but a complex system that is to some extent self-driven and can determine ends on its own. It is not always the case that we first clearly and explicitly define ends, for which we then, in a second step, seek and develop technological means. Rather, there are often inner-technological ends and driving forces leading to further technological developments that we are then asked to use.[4] In this respect, one may say that technology can become "an end in itself" (Ellul 1980: 125).

In summary, we have at least two sources that drive the technological meta-structure and that both lead to an ongoing expansion and increase of that structure: the external influence of science *and* the independent inner driving forces of technology itself. Therefore, there is no simple link between science and technology as suggested by Bacon, but rather a difficult mutual relation (Fig. 7.3.). In particular, we have a mutual impact of distinguished driving forces, which leads to an increase of the interrelated technological-scientific complex as a whole.

A final remark on some problems of the Baconian view of science and technology: Concerning the dynamics of technology, it is worth remembering the specific epistemological assumptions behind the dominant modern view of science. Because

[4] An example would be the technological development of modern cars, which entail more and more technological features. This development seems not to be driven by specific wishes and ends of the consumers, which would require respective means. Rather, it seems to be driven by the technical sphere itself, which autonomously proceeds with the development of new technological possibilities, and then asks the consumer to use them.

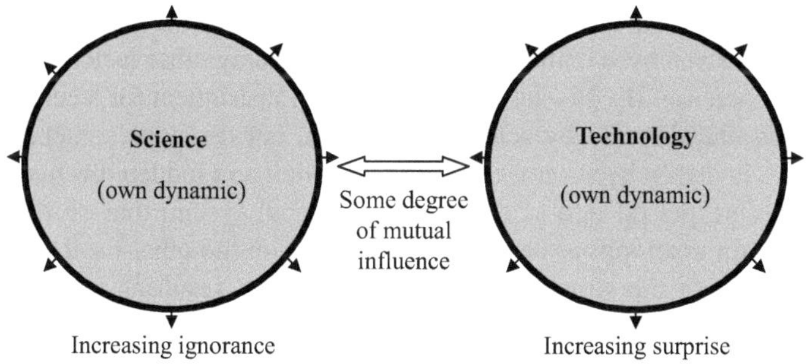

Fig. 7.3 The technology-science relation

of the linkage of technology and science, these assumptions become crucial for technology. The whole Baconian vision works on the assumption that we will reduce ignorance by scientific research, and that, therefore, technology will help to increasingly control nature. There would be, in principle, a limited set of technologies that could guarantee a perfect and complete control of nature. We are in fact confronted with the ongoing *growth* of the sphere of technology, and there is still the promise of salvation in the public awareness that all this will lead to more control of nature (e.g., genetic engineering)—that we just need more and better technology to overcome all problems we have.

However, it is doubtful that we could perfectly reach that goal and, moreover, that it would be desirable. This is for three reasons. First, the epistemological assumption underlying the meta-structure science may not hold. It may not be possible to reduce ignorance more and more. We may, rather, increase it by scientific research. In this case, there is no limited set of technological applications to govern the world. Second, there is a miscomprehension concerning the subject–object relation. It is not possible to get unambiguous objective knowledge of the world by empirical research. Knowledge is always a mental construction, designed by humans in an attempt to understand the world. Its power of explanation can be proven in the world, but there is no ultimate knowledge that fully could determine nature.[5] Because nature is always more, and always different, than every scientific theory suggests, there always remains a gap between theory and reality, a potential to surprise us. By necessity, there always will be surprises, so that nature does not always do what theory wants or expects it to do. In other words, what is *ignorance* in the field of theory—in science—is *surprise* in the field of practice—in technology. Not only may ignorance increase through the increase of science, but surprise will increase through the increase of technology (Fig. 7.3). Third, as I have argued above, there

[5]This point has prominently been made by Kant ([1781/87]1990), who argued that it is human reason that brings certain structures into the process of recognition, and thus, recognition is contingent on these structures by necessity. The human mind can only recognize objects as they are given in the structures of the mind, not as they are by themselves.

are many other internal and external driving forces, so that the technological system is not merely driven by scientific knowledge, but also by other factors that foster its dynamic of increase. By this, technology is not an instrument for a certain end that is controlled and directed by scientific research, but develops somehow "uncontrolled," i.e., is driven by several other often implicit and hidden mechanisms.

Thus, we are placed in a scientific-technological system that on the one hand imposes certain assumptions and values on us and on the other hand simplifies the real complexity of the structures and their dynamics, resulting in several hidden driving forces and "oughts" (ends, goals, norms, etc.) within these systems. Both lead to problems concerning the sustainability relations.

7.3 The Economy as a Meta-structure

Intuitively, most people may agree that the modern economy today is a dominant force influencing individuals and societies all around the world. The modern economy has a substantial impact on individual life, social interactions, political debates, intercultural relations, nature, etc. However, it is not easy to define what is meant by "modern economy," and what exactly its influence is. I hold that with the concept *economy* we do not denote a concrete phenomenon, an area of thought, or a clear defined area of action. It is rather a complex structure of basic assumptions, evaluations, and institutions working together, which can best be described and analyzed as a meta-structure.

The whole process and design of the modern economy is based on a set of basic assumptions and evaluations that tacitly and implicitly form its background. These assumptions generally stem from different sources within liberalism and modern economic thought. The most important ***basic assumptions*** are two about the rationality of the sphere of the economy. It is assumed that the economic processes and results are determined by the interplay of (1) a specific kind of *individual rationality*, and (2) a specific *systemic rationality*.

1. The first basic assumption is that the individual decides and acts as a selfish rational utility maximizer, i.e., as a *homo economicus*—an economic person. This means that the individual is guided in her decisions and actions by a rational reflection about her own preferences and a rational striving for maximal personal utility or gain. Individual economic rationality, which is at work here, is an instrumental and formal type of rationality. It is defined as the individual's ability to fully and consistently order all her desires and calculate their optimal realization under given constraints (see, e.g., Mas-Colell et al. 1995: 6ff). This economic rationality, together with individual wants and preferences, supposedly determine all decisions and actions of the individual person. Moreover, the individual is assumed to be selfish in the sense that she focuses merely on her own preferences and her own utility or gain in all her decisions (see Mas-Colell et al. 1995: 6ff). Or, as already Adam Smith prominently put it: individual self-interest,

and not benevolence, is the motivation of all economic actions (Smith [1776]2000, i.ii.2). These assumptions about the individual's rationality have been the main theoretical basis of modern economic thought up to today.[6] At the same time, they have been very influential for the real economy, for the political design of its rules, for practical actions within business, etc. For instance, economically trained managers usually recognize other actors as selfish utility maximizers and the interaction with others as a rational strategic game; recommendations of economists on politics—e.g., the influential advice of economists such as Milton Friedman on the politics of Thatcher and Reagan in the 1980s—are crucially based on the assumption that people are acting as *homines economici.*

2. The second basic assumption underlying the modern economy is that in many cases social structures and their specific design determine the overall outcome of social interactions. It is largely a *systemic rationality* that determines the overall results of social interactions, not the individual intentions of persons involved. Systemic rationality means that the structural determination follows certain logical laws and, thus, is calculable and predictable. The most prominent social structure characterized by systemic rationality is the market. In the market everyone follows her own interests and optimizes her own utility and, by the very design of the market, the overall social outcome systematically will be the best, regardless of whether this was the intention of any of the individuals acting within the market system. Adam Smith originally recognized this systemic function of the market in 1776:

> Every individual necessarily labors to render the annual revenue of the society as great as he can. He generally, indeed, neither intends to promote the public interest, nor knows how much he is promoting it. [… B]y directing [his] industry in such a manner as its produce may be of the greatest value, he intents only his own gain, and he is in this, as in many other cases, led by an invisible hand to promote an end which was no part of his intention. […] By pursuing his own interest he frequently promotes that of the society more effectually than when he really intends to promote it. (Smith [1776]2000, iv.ii.6)

The market coordinates individual interactions in a systematic way and leads to a socially optimal result. Adam Smith referred in a rather speculative way to this logic of the market as the result of an invisible hand. Modern economics, in contrast, has established a theoretical framework in which this systematic

[6] I here restrict my considerations to the established standard definition of economic rationality. Actually, economic rationality has been a major methodological issue in modern economics, and there are many variations, extensions, and enlargements of it. Several Nobel prize winners prominently dealt with a reinterpretation or redefinition of economic rationality, for instance, Herbert A. Simon (1945, 1978), Gary S. Becker (1976), Amartya Sen (1987, 1993, 2002), and Daniel Kahneman (2003). Recently, there has also been a growing number of ad hoc assumptions and variations of rationality assumptions within the research fields of behavioural economics and experimental economics. For a discussion of recent developments, see Hargreaves-Heap (1989). For a more general discussion of the homo economicus concept, see Manstetten (2000) and Kirchgässner (1991).

function of the market can be analytically deduced. As a result, the first fundamental theorem of general equilibrium theory states that, under certain assumptions, market equilibria are Pareto-efficient (Arrow 1951; Debreu 1959). This formal version of the invisible-hand function of markets is based on the assumption of individual economic rationality, i.e., that all individuals act as homines economici, and on a specific definition of the socially desirable outcome as a Pareto-efficient state, i.e., a state in which no one can be made better off without making some other persons worse off (Mas-Colell et al. 1995: 307).

The systemic rationality of an invisible-hand process is not just a characteristic of the market. In modern terms we speak of an *invisible-hand process* in a general sense, if the result of a social interaction is necessarily and systematically determined by the design of its structural framework, and not by the intentions of the individuals (Hayek 1967). This can be found in several patterns of social interactions. A rather simple example is the following invisible-hand process: if we assume a certain setting of buildings and sidewalks on the campus, there will occur over time dirt tracks on the lawns, although no one individually has intended them. This is because everyone chooses the shortest ways between the buildings, and given the case that this is across the lawn, over time a dirt track will result along shortest distance line. The common motivation and rationale of the individuals is to get from one building to another as fast as possible; the result is the track. The track and its exact direction is a systematic result that depends on the structural setting: the location of the buildings, lawns, and official sidewalks between (Keller 1990: 99ff).[7]

Systemic rationality is also assumed in game theory, which, besides general equilibrium theory, is the second main theoretical basis of modern economics. Game theory identifies and analyses structural settings of human interactions. It analyzes human interactions as strategic interactions in which each individual determines her own decisions and actions by reflecting on the overall possibilities for action of all persons involved and all the potential consequences of these actions. Game theory focuses on the systematic rationality of different settings of strategic interactions. It attempts to demonstrate that by the very design of the setting, the result of the game, i.e., the result of the strategic interaction situation, can be calculated and predicted. Again, this is not merely a theoretical approach, but has been of great influence on the real interaction among firms (see, e.g., Tirole 1988).

It is one of the merits of modern economics to have recognized the importance of social structures for human interactions and to have developed a theoretical framework for their analysis (Hayek 1967). However, this is a specific perspective on the issue based on very strict theoretical assumptions about the human being

[7] Another example would be a traffic jam that occurs at the rush hour (Keller 1990: 99f). If there is a certain density of traffic, the braking of one car systematically results in a traffic jam behind. Presumably no individual intended to cause the traffic jam. The individual intensions are just to avoid crashing into the next car. However, just by the given characteristics of the traffic—density, speed, distance of the cars, etc.—the result occurs necessarily and systematically. It can be calculated and predicted by computer simulations.

and human interactions. It easily may result in the vision that human action and interaction is *completely* calculable and predictable and could be *fully* organized and directed by the design of social structures. It may result in the idea that all political, social, and even ethical issues can be managed by an adequate structural design of the framework and rules of the individual's interactions (see, e.g., Homann 2002). However, economics offers rather limited methods for the analysis of the ethical meaning of social structures and their relevance for individual morality. In particular, they are inappropriate for the analysis of the meta-structures and the issue of sustainability. It is, for instance, not possible to analyze within the economic framework its *own* ethical relevance for sustainability.

There are several important ***basic evaluations*** underlying the modern economy, some of which are related to modern economic theory, although they are not direct derivations. I will address four crucial basic evaluations. (1) First, the economic understanding of the human being as *homo economicus* is not a mere descriptive concept, but has several normative connotations: Selfishness is no longer seen as a negative trait of humans, but rather as a positive motivation of action, because of the idea that selfish actions lead to positive results for society if embedded in adequate structures, such as the market. Similarly, economic rationality also has a prescriptive meaning, as it is widely held that one must act rationally within both business and life, to maximize one's own utility or gain. Not to do so may usually be judged as foolish. Moreover, there is a certain claim of totality in regard to the concept of *homo economicus*. The concept is not just applied to the sphere of business, but is said to be a general adequate description of the human being and her decision-making processes: Human life is in total rational maximization of individual utility (Becker 1976), and human interaction is in total a sum of strategic games. At least, it is held that the homo economicus is the only sensible and fruitful way to describe human beings and to analyze their decisions and interactions (Friedman 1953; Suchanek 2001; Homann 2002). (2) Second, there is a norm already incorporated in economic theory, namely *Pareto efficiency*. This norm enables one to distinguish a certain social state from other states, and economics is particularly interested in identifying structural conditions for a Pareto-efficient state. In public and political discussions, however, efficiency is often seen as the *only relevant* or even *ultimate* norm, and becomes a dominant guiding principle for political actions. (3) Third, an important basic assumption underlying the modern economy is *non-satiation*. In economic theory, non-satiation is a rather technical but important assumption on human preferences: it is assumed that preference relations have no satiation point. In fact that means that individuals prefer to have more of at least one good (see Mas-Colell et al. 1995: 42f and 549ff). In practice, this assumption is heavily extended to a general normative *paradigm of non-satiation and growth*. More production and goods are assumed to be better than less. On the individual level this is considered to lead to more happiness and can result in an attitude of consumerism—i.e., in the equation of consumption and individual happiness. On the social level, economic growth is widely evaluated as positive because it is tacitly assumed that this improves the overall social well-being and happiness. In short, economic growth becomes an important aim and value for individuals and societies. (4) Fourth, there is a strong connection drawn between certain liberal values and the modern economy, prominently between individual freedom and the market economy. In the modern market economy individuals interact voluntarily and agree to transactions or contracts only if it is in their own self-interest. By a voluntary economic transaction all participating actors are better off (or at least not worse off) than before. Otherwise, they would not have agreed to the transaction. In this specific way the economic system is said to be a sphere of individual freedom (Friedman 1970; see also Hayek 1994).

The main ***driving forces*** of the economy are the underlying basic evaluations, particularly the paradigm of growth and non-satiation. The underlying evaluation that more is better than less is actualized in many ways: in the social status and appraisal of richness; the equation of richness, success, and happiness; and the equation of economic growth and national welfare. However, it is not just the internalization of these values in the minds of the individuals that drives the modern economy, but also its support by powerful interest groups within business and societies as well as by the economically leading nations of the world. Moreover, the various implementations of these values in everyday life, and several institutions and organizations, lead to an actualization of them and to an automatization of its drive for the meta-structure economy. It is, therefore, important to recognize and analyze the mutual interplay of the theoretical evaluations and the institutionalization of the modern economy, and how they work together to drive the modern economy.

The ***institutionalization*** of the modern economy is realized by a complex set of institutions and organizations. Two prominent and important institutions are the market and private property. Neither are naturally given phenomena, but rather are originally specific mental constructs that have to be established and realized by a set of rules, national laws, and international agreements. Only with these rules can the market, for instance, become a practical relevant institution. Rules and laws aim to realize the theoretical conditions that are necessary for the market to be efficient, such as, e.g., free access, proper information, or proper competition. For instance, collusion on prices is not allowed, nor is a monopoly. For this, we find national laws in most countries, such as the antitrust law in the United States or the Gesetz gegen Wettbewerbsbeschränkungen (GWB) in Germany. Moreover, there are organizations to control and realize these rules, such as the Federal Trade Commission in the United States, or the World Trade Organization (WTO) on the international level. That means that many efforts are necessary to design and establish the idea of the market as a real efficient coordination mechanism on the national and global level. In addition, we have various other institutions and organizations within the economy, e.g., corporations and banks, that contribute to establish and foster the idea of the market economy. The whole institutionalization of the economy does not just result in the realization and actualization of its basic assumptions and evaluations, but also contributes to the drive of the structure and gives it an additional momentum of inner dynamic. For instance, direct individual incentives within firms and corporations to increase revenue, such as revenue–dependent payments, result in the establishment of the relation between success and growth, and implement a concrete inner driving force.

In summary, the modern economy is a complex meta-structure composed of basic assumptions about individual and systemic economic rationality, basic evaluations such as the growth paradigm, a complex institutionalization which implement and actualize these assumptions in reality, and an inner driving force essentially based on the paradigm of growth and non-satiation. Important consequences of the interplay of these elements are the inherent dynamics of increase and increasing complexity of the modern economy, which generally result from its growth and non-satiation paradigm, and the increasing influence of the modern economy on human and social life, which is particularly enforced by the claim of totality of its basic assumptions. Together, theoretical assumptions, practical institutionalization, and driving forces give the modern meta-structure of the economy its power over both our actions and thoughts on the individual, social, and global level; together, they cause the specific impact of the modern economy on the way we understand and actualize our relations with other contemporaries around the world, with future generations, and with nature.

7.4 Interrelations Among Meta-structures

I have analyzed so far the elements and, in particular, the inner driving forces of three important meta-structures. All three are particularly characterized by certain dynamics of increase. However, as I have already discussed in the case of science and technology, the dynamics of each meta-structure is not solely determined by its own driving forces, but is also influenced by other meta-structures. There is indeed a strong interconnection and mutual interrelation among the meta-structures. We are not faced with isolated and separated structures, but rather with a complex interrelated *web of meta-structures*. In this web of meta-structures, the driving forces of each structure also affect the dynamics of other structures. This results in a complex influence on the dynamics of each of the meta-structures, as well as on the web of meta-structures as a whole.

The meta-structure science is not only driven by the inner-scientific forces analyzed above, but rather is also heavily influenced by the spheres of technology and economy. As discussed above, technology is driven by science, but it also has its own inner dynamic. There are improvements and inventions of technology independent of science, occurring by coincidence, by creativity, by know-how of acting persons, path-dependency, and so on. To some extent, scientific research is driven by such technological inventions, e.g., by new technological possibilities of observation.

Moreover, both science and technology are driven by the economy. Science and technology are influenced by the potential of scientific insights and technological developments to lead to new products that can be successfully introduced into the markets. Scientific research and technological development are, in many cases, oriented toward generating economically useful outcomes. On the other hand, the economy is influenced by science and technology, and its results. New scientific insights and technological possibilities give new stimulus and options for economic production and goods.

Therefore, the driving forces do not influence merely the meta-structure they were part of, but other structures as well. For instance, the striving for an increase of utility,

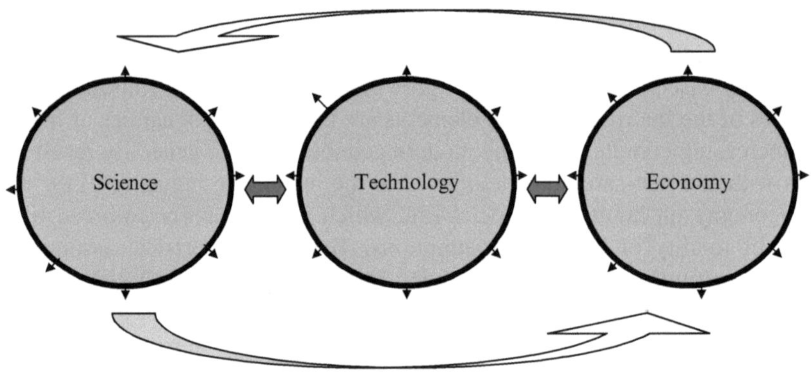

Fig. 7.4 The interrelated web of meta-structures

welfare, and gain, which drives the economy, has an effect on the meta-structures science and technology by increasing demand for scientific research for new products and technological innovations. With this, the driving forces of the economy also drive science and technology.

Furthermore, the basic evaluations of the different meta-structures are also interrelated. The striving for more goods parallels the striving for more scientific knowledge, and the striving for more knowledge goes hand in hand with the striving for more practical application of knowledge. Also, the ideal of independence and autonomy of the individual is the underlying paradigm of modern science, technology, and the modern economy. All three meta-structures support and realize this paradigm. There is also a strong interrelation of two basic assumptions: The modern scientific worldview, with its subject-based theory of knowledge and rationality, and the economic concept of individual rationality. Both are specific expressions of modern European thought and its concepts of the human being and human rationality.

Overall, the meta-structures should be understood as an interrelated web (see Fig. 7.4) whose dynamics result from the interplay of the driving forces of the different structures. The basic assumptions and evaluations, although they are specific and distinct expressions, have some parallels resulting from their common ground in modern European thought.

References

Aristotle. (2000). *Nicomachean ethics* (R. Crisp, Ed. & Trans.). New York: Cambridge University Press.

Arrow, K. J. (1951). An extension of the basic theorems of classical welfare economics. In J. Neyman (Ed.), *Proceedings of the second Berkeley symposium on mathematical statistics and probability* (pp. 507–532). Berkeley: University of California Press.

Bacon, F. [1620](1863). *The new organon* (J. Spedding et al., Trans.). Boston: Taggard & Thompson.

Becker, G. S. (1976). *The economic approach to human behaviour*. Chicago: The University of Chicago Press.
Becker, C., & Manstetten, R. (2004). Nature as a you. Novalis' philosophical thought and the modern ecological crisis. *Environmental Values, 13*, 101–118.
de Solla Price, D. J. (1963). *Little science, big science*. New York: Columbia University Press.
Debreu, G. (1959). *Theory of value. An axiomatic analysis of economic equilibrium*. New York: Wiley.
Ellul, J. (1980). *The technological system*. New York: Continuum.
Friedman, M. (1953). The methodology of positive economics. In M. Friedman (Ed.), *Essays in positive economics* (pp. 3–43). Chicago: University of Chicago Press.
Friedman, M. (1970, September 13). The social responsibility of business is to increase its profits. *New York Times Magazine*, pp. 32–33, 122–126.
Hargreaves-Heap, S. (1989). *Rationality in economics*. Oxford: Blackwell.
Heidegger, M. (1950). Die Zeit des Weltbildes. In M. Heidegger (Ed.), *Holzwege* (pp. 75–114). Frankfurt: Klostermann.
Heidegger, M. (1962). *Die Technik und Die Kehre*. Stuttgart: Neske.
Heidegger, M. (1977). *The question concerning technology and other essays*. New York: Harper & Row.
Heilbroner, R. L. (1967). Do machines make history? *Technology and Culture, 8*, 335–345.
Homann, K. (2002). *Vorteile und Anreize. Zur Grundlegung einer Ethik der Zukunft*. Tübingen: Mohr.
Kahneman, D. (2003). Maps of bounded rationality: Psychology for behavioral economics. *The American Economic Review, 93*, 1449–1475.
Kant, I. [1781/87](1990). *Kritik der reinen Vernunft* (3rd ed.). Hamburg: Meiner.
Keller, R. (1990). *Sprachwandel. Von der unsichtbaren Hand in der Sprache*. Tübingen: Francke.
Kirchgässner, G. (1991). *Homo oeconomicus. Das ökonomische Modell individuellen Verhaltens und seine Anwendung in den Wirtschafts- und Sozialwissenschaften*. Tübingen: Mohr.
Manstetten, R. (2000). *Das Menschenbild der Ökonomie. Der homo oeconomicus und die Anthropologie von Adam Smith*. Freiburg: Alber.
Mas-Colell, A., Whinston, M. D., & Green, J. R. (1995). *Microeconomic theory*. Oxford: Oxford University Press.
Mittelstraß, J. (1998). *Die Häuser des Wissens. Wissenschaftstheoretische Studien*. Frankfurt: Suhrkamp.
Mittelstraß, J. (2001). *Wissen und Grenzen. Philosophische Studien*. Frankfurt: Suhrkamp.
Sen, A. (1987). *On ethics and economics*. Oxford: Blackwell.
Sen, A. (1993). Capability and well-being. In M. Nussbaum & A. Sen (Eds.), *The quality of life* (pp. 30–53). Oxford: Oxford University Press.
Sen, A. (2002). *Rationality and freedom*. Cambridge/London: Harvard University Press.
Simon, H. A. (1945). *Administrative behaviour*. New York: Macmillan.
Simon, H. A. (1978). Rationality as process and as product of thought. *The American Economic Review, 68*, 1–16.
Smith, A. [1776](2000). *The wealth of nations*. New York: The Modern Library.
Stuhlhofer, F. (1983). Unser Wissen verdoppelt sich alle 100 Jahre. Grundlegung einer "Wissensmessung". *Berichte zur Wissenschaftsgeschichte, 6*, 169–193.
Suchanek, A. (2001). *Ökonomische Ethik*. Tübingen: Mohr.
Tirole, J. (1988). *The theory of industrial organization*. Cambridge: MIT Press.
United Nations Environment Program (UNEP). (2007). *Global Environment Outlook GEO-4. Environment for development*. Malta: Progress Press.
von Hayek, F. A. (1967). The results of human action but not of human design. In F. A. von Hayek (Ed.), *Studies in philosophy, politics, and economics* (pp. 96–105). Chicago: The University of Chicago Press.
von Hayek, F. A. (1994). Grundsätze einer liberalen Gesellschaftsordnung. In F. A. von Hayek (Ed.), *Freiburger Studien. Gesammelte Aufsätze* (2nd ed., pp. 108–125). Tübingen: Mohr Siebeck.

Chapter 8
The Impact of the Web of Meta-structures on the Sustainability Relations

The individual, as well as societies, is located in a complex web of meta-structures. Individual and societal life, thought, and actions are heavily affected and governed by these structures. In particular, the meta-structures play a crucial role for the understanding and actualization of the sustainability relations. It is not only each individual meta-structure that has a distinct impact on the sustainability relations, but the complex, interrelated web of meta-structures as a whole. By the mechanisms of the web of meta-structures we are already set in a certain way into the sustainability relations. The individual cannot choose and realize these relations with absolute autonomy or in isolation. Rather, the realization and actualization of the sustainability relations is already determined to a large extent by the meta-structures. Moreover, the individual does, by her very living and acting, participate in the meta-structures, and with this, participate in the specific way the meta-structures affect and govern the sustainability relations.

Given the complexity of the whole web of meta-structures, it would be very difficult for each individual to identify all mechanisms in which she is located and participating, or all specific resulting impacts of these on the sustainability relations. We are always set in certain ways into the sustainability relations and, by this, specifically related to other humans, future generations, and nature, but we are in most cases not fully aware of that fact. For instance, modern societies and individual lives are to a high degree governed by technology. Human relationships are strongly defined by technological means. We often relate to each other by using communication tools, such as phone or email, and approach and meet other people by using transportation means, such as car, train, and aircraft. All this requires a web of infrastructure and institutions. Email communication, e.g., requires computers, providers who set up the infrastructure of cables and servers, electrical power and its attendant infrastructure, and so on. We are all located in these patterns of action, using these technologies, which have become part of our social lives, and which are essential for private and business relations. Thus, most of our relationships are technologically mediated. Moreover, these technologies set us in many tacit relations with people around the world, as well as with nature and future generations, because they require energy, resources, result in e-waste, etc. This affects people in other

C.U. Becker, *Sustainability Ethics and Sustainability Research*,
DOI 10.1007/978-94-007-2285-9_8, © Springer Science+Business Media B.V. 2012

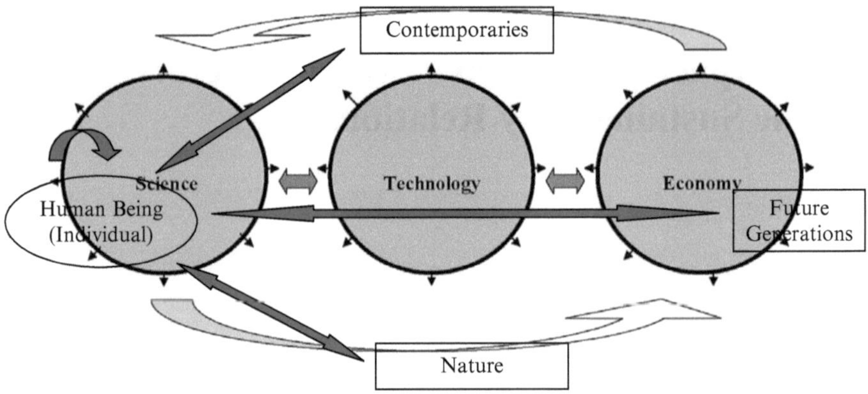

Fig. 8.1 The impact of meta-structures on the sustainability relations

parts of the world, future generations, and nature in complex ways, many of which we are not aware and are not able to fully identify.

There are indeed many hidden, implicit, and asymmetrical forms of relation between humans set up by technology and the modern global economy. In business, technological specialization, in combination with transportation and the market economy, has led to specific indirect and tacit relations between consumers and producers. By using simple products such as clothing, food, or a cellular phone, we take part in very complex global processes of production, distribution, consumption, and waste management, and are related in many ways to a rather huge number of people around the world who are participating in these processes. Intercultural relations are set up in specific ways, supported and shaped by business and by technology. For instance, intercultural contacts are often reduced to mass tourism—enabled by modern transportation means—or to an indirect contact through television and mass media—enabled by modern satellite technology. Finally, our relation with nature is to a large extent mediated by modern economy and technology: Modern food production and agriculture is heavily defined and structured by technology, we have air conditioning and heat to protect us against weather conditions, we use technology in medicine, etc. Even when we explicitly approach nature as wilderness, we use outdoor equipment.

We may say that the meta-structures work like filters through which we are related to ourselves, other humans, and nature (Fig. 8.1). Their underlying basic assumptions, evaluations, driving forces, and institutionalization all heavily influence our sustainability relations. The meta-structures form a complex interrelated web that causes a specific understanding, realization, and actualization of the sustainability relations.

At this stage of the analysis, the difficulty of the structural issue of sustainability ethics becomes evident. It is not just an issue of one single ethically relevant institution or pattern, but a complex web of interrelated meta-structures that should be taken into account. The interrelation of the meta-structures, and the mutual influence of their elements, causes a complex and difficult setting of patterns, values and driving forces in which the individuals are located. This affects the potential of individual morality and its actualization in regard to the sustainability relations. The structural challenge of sustainability ethics is to analyze the ethical meaning of this impact in detail; to analyze the ethical meaning of the meta-structures and their impact on the individual and her sustainability relations; and to analyze the compatibility of the structural dimension with an ideal realization of the sustainability relations. It should particularly be taken into account that the meta-structures show a dynamic of increase that is driven by a complex mutual interaction of their driving forces. By this, they have an increasing influence on individual thought and action. The meta-structures influence thought and action, as well as the sustainability relations, more and more and, thus, have an increasing impact on the ethical aspects of these relations.

To analyze possibilities for a more sustainable organization of life, societies, and global future, an encompassing approach is necessary, which particularly includes an explicit ethical analysis of the influence of the meta-structures and their dynamics on human life and relations. Problematic elements have to be identified, and a potential redesign of the meta-structures has to be suggested.

Part III
Toward a New Sustainability Ethics

The first two parts of this book gave a determination of the issue of sustainability and of the challenges of sustainability ethics. I have focused so far on two main challenges—the relational and the structural challenge. In Part II, the importance and the difficulties of the structural aspects of sustainability were emphasized and analyzed in detail. It might appear that the individual and her sustainability relations are essentially governed by the meta-structures and that, therefore, the ethical issue of sustainability is merely a structural issue of the ideal design of meta-structures. However, this is not the case. Sustainability ethics is not merely a structural ethics. The structural aspect is only one aspect. The meta-structures are indeed a necessary and crucial element for the establishment of the sustainability relations. However, the sustainability relations are not completely constituted by them. They are rather developed by the interplay of individual actions and attitudes *together* with the meta-structures. Sustainability ethics should analyze both the individual person and her role within the sustainability relations, and the impact of the meta-structures. We have to analyze the implications of the relational dimension of sustainability—of the embedding of the individual person into the sustainability relations—for the individual person and individual morality; and we have to analyze the way the individual and its sustainability relations are affected by the meta-structures. I will first focus on the person (Chap. 9) and then on the meta-structures (Chap. 10). The original point of reference is the sustainability relations and their characteristics. I therefore proceed by returning to the analysis of the relational aspects of sustainability defined in Chap. 3 and Sect. 4.3, and developing some fundamentals of sustainability ethics in regard to the individual person and her morality in the following Chap. 9. This will in a second step enable us to address the structural dimension of sustainability ethics in Chap. 10 and to give an ethical evaluation of the existing design of meta-structures, as well as to develop ethical guidelines for their redesign.

Chapter 9
The Relational Dimension of Sustainability Ethics and the Role of Individual Morality

The concept of sustainability is a concept about continuance, relationships, and orientation. As such, it has an ethical dimension. The ethics of sustainability deals with questions of how one should live within the sustainability relations – within the threefold relatedness with contemporaries, future generations, and nature. Sustainability ethics cannot be developed independently or detached from the sustainability relations. Sustainability refers to the human being as a relational being, and the ethical issue of sustainability can be addressed only with regard to this fact. An approach that first abstracts the person from her sustainability relations to deduce an ethics, and then applies it in a second step to the issue of sustainability, is misleading. A sustainability ethics must from its very beginning be developed by analyzing the individual as a relational person within the contexts of the sustainability relations.

This chapter discusses some basic issues of the relational dimension of sustainability ethics. For this, I go back to the sustainability relations and proceed with the analysis of their characteristics and ethical implications. I argue that one crucial fundamental of sustainability ethics is the identity and self-understanding of the person as a *relational*, *interdependent*, and *virtuous* person in the context of the sustainability relations, i.e., as a *sustainable person*. This is based on a broad integrative understanding of the human being as an emotional, rational, creative, and communicative being. However, the development of the personal identity and the relationships are mutually dependent on each other. The sustainable person and the sustainability relations are interrelated and develop simultaneously in a dynamic process, and the excellence of the person and of the relations necessarily belong together. This is no static concept: there is neither a "perfect" sustainable person, nor a "perfect" sustainability relation. It is an ethics of development and perfection, not an ethics of fixed rules. It is a procedural ethics of an ongoing development of both. Personal identity is constituted by and simultaneously constitutes the sustainability relations, and vice versa. This also is not a merely theoretical concept, but is grounded in and actualized by the concrete details of individual experience within the context of the sustainability relationships.

C.U. Becker, *Sustainability Ethics and Sustainability Research*,
DOI 10.1007/978-94-007-2285-9_9, © Springer Science+Business Media B.V. 2012

This means to consider both the person and her relationships together and to analyze the characteristics, potentials, and requirements of each of the relationships, as well as the respective abilities and competencies of the person to adequately recognize and respond to them. In the following, I discuss some main features of the sustainability relations and the sustainable person. I carefully distinguish the differences of each of the sustainability relations and finally identify a set of excellent personal abilities and competencies, the integration of which constitute the sustainable person. I start with the human–nature relationship, which is the most difficult one because of its distinctiveness to relationships between humans, and because of the dominance of existing scientific, technological, economic and gender patterns affecting it.

9.1 Ethics of the Human–Nature Relationship

What are the crucial characteristics of the human–nature relationship, and what are the ethical implications of the specifics of this relationship? I want to start with a general definition of "nature": nature is that area of reality that comes into being and exists independently of human thought and action. That I define nature by using the concept of human thought and action is not a mistake, but rather a necessity. As a fundamental concept, nature is defined in relation to the human being and vice versa. The self-understanding and identity of the human being can be defined only in relation to nature, as the ultimate, given condition of human thought and action. This does not mean that I want to establish a simple dualism or complete separation of human and nature. I rather want to maintain a fundamental mutual relationship between nature and human beings, which constitutes a necessary condition of the being of both. This even would allow for the idea of an ultimate unity of both.

It all depends on the further interpretation of the relationship. One can stress either the otherness or the sameness—the dissimilarity or the similarity—of both. Extreme positions include, on the one hand, positions such as those incorporated in scientific and economic patterns of thought and action, which stress the otherness of humans and nature by referring to certain concepts of rationality as the main distinction and the main means of encounter; and, on the other hand, positions that stress the sameness of humans and nature—their identity or even unity—such as deep ecology or Darwinism (see also Becker and Manstetten 2004). The former establish a strict separation and dualism, which ignores aspects of sameness between nature and humans. The latter positions focus on the sameness of human and nature, resulting in a disregard of the distinction between, and the otherness of, both. Such extreme positions have been very influential in the perception of and discussion about the human–nature relationship. However, they result in simplifications of that relationship, and in a limited recognition of its ethical aspects.

I hold that an appropriate understanding of the human–nature relationship requires one to carefully consider both the otherness *and* the sameness of humans and nature. This will allow us to fully recognize all characteristics of the encounter, correlation, and integration of both, and to adequately determine the ethical aspects of their relationship. For this, it is important to identify all ways by which we are related to nature. We are not merely related to nature by rationality, by scientific or economic approaches to nature. We are also related to nature by emotions, biological necessities, personal experience, creativity, etc. This has been particularly analyzed on a philosophical level by the Romantics. An elaborated contribution has been given, for instance, by Henry David Thoreau. He demonstrated that one can encounter nature by rational, aesthetic, emotional, contemplative, and meditative ways and, with this, recognize or experience nature in both its otherness and sameness.[1] At the same time, this allows the human being to recognize herself as a being possessing all these potentials of experience and recognition. For Thoreau, the development of an experiential and personal relationship with nature results in a deeper insight into human existence and the development of human excellence.

A further important way of relation has been stressed by several other Romantics, including Novalis, Schelling, and Wordsworth. They analyzed the meaning of creativity for the human–nature relationship and stressed that creativity is something humans have in common with nature.[2] Nature and the human mind are both creative and productive. They can bring forth a variety of new forms, develop entities, are regenerative, and so on. For these thinkers, the human being is indeed primarily a creative being rather than a rational being. Creativity, thus, to some extent constitutes a sameness of humans and nature and allows a relationship between them. However, creativity also constitutes a moment of otherness, as human creativity can be considered as a specific development or fulfillment of nature's creativity. This led the Romantics to posit a distinction between a destructive and a constructive form of human creativity. The destructive form is ignorant about its foundations in nature. An example for this type of creativity would be the modern economic productiveness. The constructive form recognizes and respects its origins and strives to orient and integrate human creativity into nature's creativity[3] (one concrete example of this might be the concept of an English garden).

All these considerations about the human–nature relationship identify aspects of sameness and of otherness of humans and nature, and argue for a broader idea of encounter of both on various levels. They are based on broader concepts of nature, humans, and the human–nature relationship and particularly allow for the understanding of nature as another self in its own right and not just as another object or as something that lacks rationality, etc. To recognize and approach the other as another self means both recognizing my counterpart's similarities to myself, as well as acknowledging our differences. Only if both aspects are accepted

---

[1] See particularly *Walden* (Thoreau 1992) and *Thoreau's Natural Essays* (Thoreau 2002). For a more detailed analysis of Thoreau's environmental thought, see also Becker (2003).

[2] See Becker and Manstetten (2004) for a detailed analysis of Novalis's view of this interconnection of human and nature, and Becker et al. (2005) for an analysis of Wordsworth's version.

[3] See, e.g., Wordsworth ([1814]1936: VIII 87–94, IX 1ff, XII 370ff). See also Becker et al. (2005).

may a deep relationship be developed (see Becker and Manstetten 2004). The human–nature relationship, thus, provides the potential for a fundamental experience of otherness and interdependency.

To understand nature and the human–nature relationship in this way is not a theoretical issue. We cannot fully recognize the other self of nature, its otherness and sameness, by rational analysis or theoretical reflection alone. We have to experience it in real and personal encounters. Thoreau particularly stressed this point. He recognized that a mere theoretical definition would mean a determination of nature and its reduction to a simple concept. Only personal encounter by rational, aesthetic, emotional, contemplative, and meditative ways allows for the development of an attentiveness that would acknowledge both the otherness and sameness of nature—its other being and self—and for a fully developed relationship with it.[4] The human–nature relationship is not something that can be defined merely theoretically or determined by (scientific or economic) rationality. It is something in which the human being is necessarily located, and it can only fully be understood by living within this relationship and developing personal attitudes of attentiveness and openness to the sameness and otherness as well as to the various levels of mutual encounter.

In terms of virtue ethics (which, recall, is concerned with the development of the virtuous person), the ethical issue concerning the human–nature relationship is the development of certain virtues—*environmental virtues*. I define *attentiveness* to and *receptiveness* toward nature as basic environmental virtues. They allow for recognizing and experiencing the otherness and sameness of nature as well as its other self, and are a prerequisite for the development of other environmental virtues, such as an attitude of respect for nature. These environmental virtues can not be developed in other contexts and then be applied to nature. They are not general virtues of attentiveness and receptiveness, which the human being can train elsewhere, e.g., in a social context in relation to humans, and then apply to nature. These virtues are specific kinds of attentiveness and receptiveness toward nature and can be developed only through interaction and encounter with nature. However, we can also identify other specific kinds of receptiveness and attentiveness as crucial excellent attitudes in the context of the other sustainability relations. Therefore, and because attentiveness and receptiveness are more fundamental attitudes and prerequisites for the development of further virtues, I have labeled them *basic environmental virtues*, and will also call them later on *relational competences*.

---

[4] Thoreau did not primarily strive for an analytical treatise on nature in his works. He did not want to solve the question of the relationship between humankind and nature theoretically nor possibly to provide a final answer. "[…T]he meaning of Nature was never attempted to be defined by him" (Emerson [1862]1998). In *Walden*, Thoreau carefully placed all levels of encounter with nature side by side, without any attempt to actively systematize or order them. Thoreau's style of composition aims to express the variety of the dimensions of nature and its other self, with all its potential contradictions, in a literary work (Becker 2003, 2008). In this sense, Thoreau's writing about nature indeed shows a "programmed inconsistency" (McIntosh 1974: 17).

At this point it is worth referring to the origins of virtue ethics within Aristotle's philosophy. Aristotle (2000) gave an elaborated theory of the mechanism of development of morality, i.e., virtues. For Aristotle, morality is not solely based on rationality—as, e.g., deontology suggests—nor is it solely based on emotions—as, e.g., Hume [1772](1998) suggests. It is rather the interplay between both emotions and rationality that is the basis of morality. According to Aristotle, every human being has specific inborn appetencies and emotions that originally motivate his actions. The person is, however, able to cultivate the appetencies and emotions over time through practical experience and by the assistance of practical wisdom (*phronesis*), and, with this, to transform them into virtues (Aristotle 2000: I and II, 1144b 1–30). It is important to recognize that Aristotle carefully distinguished different kinds of rationality (*logos*): forms of rationality related to theoretical thinking, e.g., *episteme*, which enables the human mind to recognize abstract knowledge such as mathematical and logical structures, and forms of rationality related to practical actions and production (Aristotle 2000: 1138b 16–1145a 10). Because ethics is concerned with rightness in the field of practical action, it cannot refer to theoretical forms of rationality, but has to refer to a practical kind of rationality—to practical wisdom. This refers to the personal and contextual ability of the human mind to determine the appropriateness of actions in concrete situations and relations. Because every situation is different and new, practical wisdom is the skill of right judgment in new situations and cases. This exists in contrast to theoretical rationality, which is concerned with unchangeable, fixed structures (Aristotle 2000: 1138b 16–1145a 10). In addition, communication—debate and consultation—within the community of the polis is another important pillar of moral foundation, i.e., of the determination and development of virtues (see also Rese 2003).

Aristotle had in mind virtues that were crucial for the specific relationships of a small ancient polis, such as courage, temperance, and generosity (Aristotle 2000: 1114b 26–1121a 15). They all can be developed from inborn appetencies or emotions. For instance, someone might be more foolhardy or fearful by disposition. In action over time, and by the judgment of practical wisdom, he or she will be able to train his or her original disposition to the virtue of courage (see Aristotle 2000: 1114b 26–1121a 15). In the context of sustainability ethics, we are interested in a rather different relationship—the human–nature relationship—and in respective environmental virtues such as attentiveness, receptiveness, and respect for nature. It is, however, possible to think about these virtues within the theoretical framework of virtue ethics if we consider certain dispositions or abilities, such as sympathy for other beings, as dispositions that are more or less innate in every human being.[5] A disposition toward sympathy can be developed

[5]Sympathy is a well discussed concept in the history of philosophy. Rousseau ([1755]1995: 70f) prominently suggested that humans have an original disposition for sympathy with all beings, human and nonhuman. Sympathy has also been crucial for the ethical considerations of Smith ([1759]2000: I.i.1.1), although he mainly was interested in the ethical meaning of sympathy for the relationship between humans.

into environmental virtues of attentiveness and respect for nature. This requires an ongoing encounter and interaction with nature guided by practical wisdom. The interaction and encounter with nature alone does not directly result in environmental virtues. It is the *reflected experience* of the encounter with nature that cultivates environmental virtues. Environmental virtues would, therefore, combine the contextuality of personal experience with the kind of rationality Aristotle has called practical wisdom.[6]

The ethics of the human–nature relation is to a large extent an ethics of the personal development of excellent attitudes toward nature, of environmental virtues, in the interaction with nature. Humans are necessarily embedded in a fundamental relation with nature, and they have the potential to develop or degenerate this relationship. The awareness and recognition of the full potential of this relationship challenges humans to strive for its realization. This is an aspect of human excellence. However, it would be a misinterpretation to understand nature as a mere means for human excellence or good life. The development of environmental virtues and, with this, of the full potential of human existence, at the same time leads to nature's excellence. It brings to light nature as another self, and respects the mystery of its being. Attentiveness to the otherness of nature ultimately is a reverence for the infinite potential of humans, nature, and their relationship.

However, the ethics of the human–nature relationship is not fully captured by the concept of environmental virtues. It also entails the recognition of the fundamental importance of this relationship—the recognition that this relationship has a value in itself in as far as it entails the fundamental potential for the development of both human's and nature's excellence—and respect and responsibility for the potential, protection, and development of this relationship. For this, we must ascribe a further kind of rationality to humans that enables the individual to generalize the personal identity and virtues she has developed by personal experience and actualization within the relationship with nature, and enables her to develop a general personal identity as a fundamentally interdependent being embedded in multiple relationships, including the human–nature relationship. Although the development of environmental virtues is a crucial basis for the ethics of the human–nature relationship, this second rational ability is necessary for further reflection on the virtues and for developing a general insight into the value of the relationship and its potential. I call

[6]As an example for environmental virtues of attentiveness and receptiveness towards non-human beings, one may think about the relation with her dog. It is possible to develop a specific attentiveness and receptiveness toward the own dog over time and through experience and personal encounter with the dog. This also enables one to recognize the specific self of the dog, and "communicate" with the dog. Such an attentiveness and receptiveness is crucial for the flourishing of the relation. It is based on a personal approach and practical wisdom, and differs from mere theoretical or scientific approaches. It is based on emotional, communicative and practical rational abilities, and not just on theoretical rationality. We could also think about developing a specific attentiveness and receptiveness towards plants, a competence, which good gardeners have.

this kind of rationality *reason*.[7] We have, therefore, a combination of two rational abilities: practical wisdom, which directly reflects on emotional motivations and relational actions, and reason, which reflects on the virtues themselves and formulates a general personal moral identity. Both abilities refer to the experience of relationships; they can fruitfully contribute to the development of morality only *within* the experience and cultivation of the human–nature relationship.

In summary, the ethics of the human–nature relationship has several constitutive basics: (1) a personal identity as a fundamentally interdependent being in its relationship to nature; (2) personal excellent competencies of attentiveness and receptiveness toward nature; (3) a (self) understanding of the human being as an emotional, rational, creative, and communicative being; (4) a broad concept of rationality, which includes practical wisdom and reason; (5) a broader understanding of human excellence, which is necessarily bound to the relationship with nature and the development of environmental virtues; with this, (6) a broader understanding of nature and nature's excellence, which also is necessarily bound on the interaction with environmentally virtuous human beings; and (7) finally, because the human–nature relationship is a necessary condition for both humans' and nature's excellence, the recognition that this relationship has a moral worth in itself.

I finally want to give two remarks on this ethical conception. First, a remark on environmental virtue ethics: In regard to the interpretation I developed above, environmental virtue ethics is not a matter of *application* of virtues to the interaction with nature, but rather a matter of *constitution* of distinguished environmental virtues related specifically to the interrelation with nature. Furthermore, it defines human (and natural) excellence in a new way, making the human–nature relationship a necessary sphere of excellence. This means that environmental virtue ethics, as I understand it, is not a simple application, but rather a substantial contribution to traditional virtue ethics theory. Second, I want to indicate that my considerations so far would particularly allow for a sensible underpinning of Aldo Leopold's Land Ethic—one of the prominent approaches in contemporary environmental ethics. Leopold [1949](1966) asked for respect for the integrity of the land community. By this he understood the community of nature as inclusive of humans. Leopold himself gave no elaborated philosophical argument for this, and later debates have questioned whether he committed a fallacy regarding the is–ought distinction (see, e.g., Callicott 1989). However, if we would interpret the land community as a relational setting, in which environmental virtues and excellence are developed by interaction of humans and nature, we could argue for respecting the integrity of this setting.[8]

[7]"Reason" here means an ability of the human mind, a kind of rationality that enables the individual to abstract from the mere personal conditions and to generalize them. It is not given, as the *Vernunft* for Kant, but has to be developed as an excellent competence. It is more like a *dianoethical* virtue, a virtue of thinking (of the logos), in Aristotelean terms. However, Aristotle's dianoethical virtue is not comparable to my definition of reason (see Aristotle 2000: Book VI).

[8]See also Shaw (1997) for a virtue ethics interpretation of the Land Ethic.

On the other hand we have to recognize that this interpretation is not fully compatible with Leopold's implicit evaluative view of the human being as a more or less external factor negatively impacting nature. Leopold seemed to think that the less human impact on nature, the better (Leopold [1949]1966: 251ff).[9] This is, of course in stark contrast to my above given considerations about the human–nature relationship and its ethics, particularly the mutually related excellence of humans and nature.

9.2 Ethical Specifics of the Relationship with Future Generations

What are the specific characteristics of the relationship with future generations, and what are the ethical implications of this relationship? First, we have to discuss how the relationship with future generations is defined and realized. There are at least two different kinds of that relationship. We can think about a direct, personal relationship with future generations, and an indirect, abstract one. The direct, personal one is that with our children and grandchildren. The indirect, abstract one is that with generations who do not live now but will be living in the more distant future. To identify the crucial aspects of our relationship with future generations, we must consider the specifics of both kinds of relationship and the respective ethical implications. One the one hand, we can identify some basic personal requirements, such as attentiveness and receptiveness, in regard to the direct relationship between parents and children. On the other hand, if we think about generations not living yet, we are confronted with a rather abstract relationship we cannot directly experience, but can only consider by rational reflection. In both cases, however, the relationship with future generations constitutes a fundamental recognition, experience, and actualization of dependency, care,

---

[9] That position also underpins the *wilderness* concept, which has been prominent in recent environmental discussions (see, e.g., Nash 2001). Wilderness is a concept that stresses the otherness of humans and nature. It maintains the dualism established by modern thought, but changes the evaluations of both parts: nature is ascribed a higher value and humans are seen as a negative impact factor. Nature as wilderness, i.e., without human influence, is better than nature that is influenced by humans. This is in contrast to my given considerations about the human–nature relationship and its ethics. If generalized and used as an ethical framework, the wilderness concept would be a reduced understanding of humans, nature, and the relationship between both. The idea of separating nature from humans would be artificial, unrealistic, and not fruitful for either. It would ignore the potential of the human-nature relation and the various ways of encounter. It would ignore, for instance, the idea of a garden, as an extreme contrast. The garden is a concept of encounter of humans and nature and in the ideal case has the potential of building human and nature's excellence. It is a mode of real encounter and interaction with nature that allows the gardener to develop attentiveness and respect for nature. One may also say it allows for care of nature. The garden does not even exclude the recognition of and the respect for the otherness of nature, because this does not depend on some objectively defined wilderness, but rather on the attentiveness of the human individual towards the other self of natural beings.

and temporality[10] of the individual, and represents the potential to develop a personal identity and excellent attitudes in regard to these aspects. In this respect, the relationship with future generations refers to our intergenerational relatedness and dependency in general, which actually always has two timely directions: past and future.

The relationship between parents and children is an elementary direct relationship between the current generation and those who will become the future generation. It allows both parties the personal experience of aspects of dependency, and asymmetry of power and abilities to communicate, think, and act. Moreover, it is a dynamic relationship. Dependency and asymmetry change, children become adults, and adults may become dependent on their children when aged. This relation forms some basic experiences and recognitions and is crucial for the development of a personal identity: We recognize and realize (self-actualize) ourselves as timely beings, both constituting and constituted by this relationship.

The characteristics of this relationship imply certain moral characteristics, e.g., responsibility, duty, and care for children, or gratitude and respect for parents. What is constitutive for this kind of morality? Is it theoretical reasoning, deduced rules, or a contract? Can one, for instance, learn by a theoretical argument or a book to be a good parent? Or do we become a good parent by experience and practice? This may be a fallacy of false alternative, but I hold that it is at least evident that a mere theoretical approach to care and education would be a shortcoming. The morality of the parent–child relationship is grounded to an important extent on the reflected experience and living of this relationship. The unique individual character of each child and the ongoing dynamic development of the relationship require an individual and flexible way of responding and acting. This again requires an excellent attitude of attentiveness and receptiveness to the child, her individual characteristic and being, and her own self. In terms of virtue ethics this means that certain virtues are required, which can only be developed by the reflected experience of the relationship with the child over time. In this interpretation, care and responsibility can be interpreted as virtues, too. They are excellent attitudes toward the asymmetric and dependency which characterize the intergenerational relationship, and can in its specifics only be developed and actualized through the relationship itself.[11]

[10]By the term *temporality* I denote the temporal conditionality of the human being—its being in time. That means both the temporally limited individual existence, as well as the outlasting of the individual existence by other beings. The experience and realization of temporality is one main characteristic of the relationship with future generations. That does not mean that it is not experienced in other contexts and relations. It is, for instance, also experienced in the relationship with nature, and it would also be possible to distinguish a relationship with future natural entities. However, for the sake of simplicity, I refer here only to the main aspects of each of the sustainability relations, and in this respect, I relate the experience and realization of temporality to the relationship with future generations.

[11]I do not want to argue that responsibility is not grounded in a process of rational reflection. However, the specific, concrete responsibility towards a child has additionally to be grounded at least in reflection on experience of the relationship itself. As such, responsibility has characteristics of a virtue. The virtue of responsibility, like other virtues, cannot merely be deduced theoretically, but must also be defined and actualized practically.

The parent–child relationship provides us with an elementary experience of the dependence and the temporality of human existence—of the intergenerational embedment of the individual. Within this relationship we can understand ourselves as beings constituting and being constituted, caring and being cared for, and we can develop excellent attitudes of attentiveness and receptiveness, care and responsibility, in regard to future (and past) generations. Therefore, the parent–child relationship is important for the timely relationship between generations, and it is a crucial constitutive element of its ethics.

However, the relationship with other generations is also a relationship with generations in the farther future or past, and is experienced and actualized in an indirect way. We experience and actualize it by several means, e.g., by transfer of knowledge, goods, artifacts, transformed nature, etc., over time. By the encounter with such things, the individual experiences herself as a timely contingent being, as well as a cultural being. Things passed from one generation to the next realize and form an indirect and asymmetric relationship between generations, and allow the individual to experience and actualize herself as a constituting and constituted being in the chain of generations. In this, it is also possible to develop an excellent attitude of attentiveness, respect, and care toward generations in the distant future or past and to develop and actualize a personal identity as a temporally interdependent and responsible being. In particular, the individual can even develop an attentiveness and respect for artifacts and knowledge, both of which are crucial for developing an excellent relationship with future and past generations. Such an attentiveness and respect recognizes the important role of artifacts and knowledge in the context of timely interdependency between generations.

I have stressed so far ways of personal experience and practice of intergenerational relations and a respective morality grounded in personal identity and virtues resulting from that experience and practice. However, this morality has to be completed by rational reflection. The ethics of intergenerational relationships can not merely refer to practical wisdom, which reflects on personal experience of these relationships. It must also include a process of reasoning by which the individual person recognizes that her identity as a temporal being, her virtues of attentiveness and receptiveness, her care and responsibility toward other generations, exceeds the sphere of mere direct personal experience, and can and should be extended to a general relationship with future and past generations. This leads to the recognition of the fundamental meaning of this relationship for human self-identity and excellence. Together, these constitute a virtuous relationship to future generations.[12]

In summary, the ethics of the relationship with future generations is grounded in the self-understanding and realization of the individual as a temporal, interdependent

[12]It might, however, be possible, to base an ethics of intergenerational relationships "merely" on personal experience and attitudes. For instance, many cultures know a direct relationship with their past ancestors and "interact" with them.

being. It develops by reflected experience and actualization of intergenerational relationships, combined with a rational reflection and extension of this identity and virtues to other generations in general. This is the basis for the development of a set of excellent attitudes in regard to future generations, such as attentiveness and receptiveness, care and responsibility. Both the virtues and the personal identity as a temporal interdependent being are crucial elements of the sustainable person and the sustainability relations.

9.3 Sustainability and the Ethics of the Relationship Between Contemporaries

Various aspects of the relationship among contemporaries may be of relevance in regard to sustainability. We may think about social interactions on the local level, which impact the local environment or the future of communities. We may think about the crucial importance of change and development of societal values and norms by the process of societal discourses. We may think about local sustainability and resilience, established by communities and societal groups.

However, I focus here on one specific dimension of the relationship between contemporaries which is prominently addressed by the modern concept of sustainability and raises particular ethical challenges: global relationships. Sustainability essentially entails the question how global living together can be designed and organized such that a good life for all people currently living on earth is possible. Global relationships are relationships between people of different regions and, in many cases, different cultures. They show some specific characteristics that are not typical for all relationships between contemporaries and distinguish them, e.g., from closer personal or local relationships.

One characteristic of global relationships is the existence of cultural differences. By this, it is a relationship between people who have different languages, concepts, knowledge, and values. Another characteristic of global relationships is the differences in the natural environments, which result in differences of the possibilities of living, creating, producing, and acting. Further, in many cases of global relationships there is an asymmetry in power and in living conditions. All these characteristics result in specific challenges of developing and realizing global relationships, and entail particular ethical challenges.

Global relationships can be developed in direct and indirect ways. One example of a direct way might be personal encounter and direct communication, while an indirect realization of the relationship might involve the exchange of artifacts, such as economic products, pieces of art, etc. Global relationships particularly allow for experiencing ourselves in a specific way as social and communicative beings, as they confront us with particular challenges. Their actualization requires us to deal with differences of concepts and values and to develop a common basis for communication.[13] Global intercultural relations require (and allow for) the recognition of the otherness as well as the sameness of different cultures.

The recognition of cultural otherness requires the encounter with other cultures and the development of an attentiveness to, and receptiveness for, that otherness. The experience and recognition of cultural otherness allows for a fuller understanding of one's own cultural contingency and determination, as well as an improved understanding of possibilities for other sets of concepts and values. This experience is a prerequisite for the development of a reflected cultural identity, intercultural competence, and the development of intercultural virtues such as tolerance and respect for other cultures.

In summary, global relationships require and allow for the development of specific kinds of attentiveness to, and receptiveness for, the otherness of other cultures. That relationship requires and allows for the development of a reflected cultural identity and intercultural competence, as well as for a specific kind of experience and actualization of the individual as a social and communicative being. This is a prerequisite for the development of intercultural virtues such as tolerance and respect.

All these aspects of personal identity, skills, and character traits are further characteristics of the sustainable person. Her cultivated intercultural competence allows for the excellence of global relationships and, with this, for human excellence, as it allows for a reflective cultural identity and respect for cultural differences.

9.4 An Integrated Ethical Approach to All Sustainability Relations: The Sustainable Person

I have argued so far that each of the three sustainability relations has its specific characteristics and, by this, specific ethical aspects. At the center is the person who develops a self-identity as a *relational person* existing in the context of the sustainability relations, as well as specific *excellent relational competences and attitudes* in regard to each of the three relations. Core concepts here are (1) a personal identity incorporating the *temporality*, *interdependency*, and *cultural contingency* of human existence, (2) the relational competences (basic relational virtues) of *attentiveness* and *receptiveness*, which are relevant for all sustainability relations but whose forms differ in each of the sustainability relationships, (3) a set of specific relational virtues, such as *respect, care, responsibility,* or *tolerance,* which are developed and actualized in different ways within one or more of the sustainability relations. This altogether requires (4) an encompassing idea of the human being as an *emotional,*

[13] In modern times, however, mass media and exchange of economic products have become the main means for an indirect realization of global relationships, as mass tourism has become for direct encounters. I will argue in Chap. 10 that this results in a problematic reduction of global and intercultural relationships.

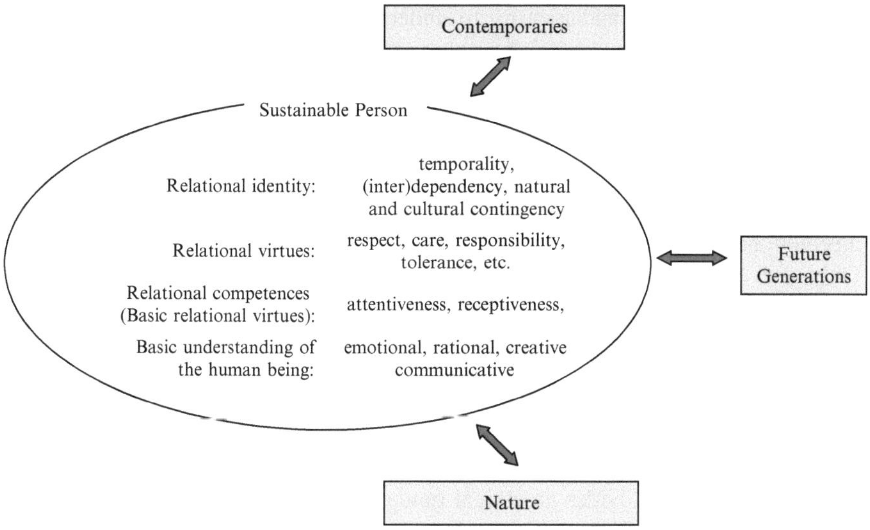

Fig. 9.1 The sustainable person in the context of the sustainability relations

rational, communicative and *creative* being, and particularly a wider concept of rationality that includes *practical wisdom* and *reason* functioning in the context of multiple relationships (see Fig. 9.1).

An encompassing sustainability ethics must bring together all three sustainability relations and their ethical aspects into one framework. If we want to fully grasp the orientational meaning of the concept of sustainability, we have to take into account the ethical aspects of all three relationships simultaneously. The ethics of sustainability is more than an environmental ethics issue or an issue of environmental or intergenerational justice, which refer only to ethical aspects of one of the sustainability relations. In its relational dimension, sustainability is an issue of all three relationships together. It is an issue of the threefold embeddedness of human existence in the sustainability relations. The main pillar of sustainability ethics, therefore, is an encompassing integrated relational identity of the person.

The sustainable person is one who recognizes and understands herself as located within all three relationships at the same time. This requires the development of a personal identity and excellent attitudes in regard to all sustainability relations. It is the reflected experience of, and actualization within, these relations that allows for the development of relational competences such as attentiveness and receptiveness, relational virtues such as respect, care, responsibility, and tolerance, and a personal identity as a relational person existing within the sustainability relations. It is important to recognize that this exceeds a traditional virtue ethics perspective, as the sustainability relations cannot completely be captured by personal experience. Personal experience is the foundation for their recognition and approach, but a further step is necessary. Based on reflected experience, the sustainable person needs to extend her competences, virtues, and identity to understand herself as existing within and intimately constituted by her relationships with nature, future generations, and contemporaries in a fundamental and general way. She needs to recognize the fundamental meaning and value of the sustainability relations as such. This finally results in a reflected integrated relational identity combined with excellent relational competences and virtues, i.e., in a *sustainable person* (Fig. 9.1).[14]

Sustainability ethics requires as its basis a certain concept of the human being as an emotional, rational, creative, communicative, and fundamentally relational being. Only such an integrative and robust understanding of the person allows for full recognition of the sustainability relations, their specific dynamics, their potentials, and their ethical implications. It follows that personal identity, competences and virtues play a crucial role for sustainability ethics, and that sustainability ethics cannot refer to a mere theoretical, rational approach. The identity, competences, and virtues of the sustainable person are the conceptual basis of sustainability ethics.

Sustainability ethics, based on the sustainability relations and the concept of a sustainable person, provides an ethical fundament for sustainability. It helps us to ask the appropriate ethical questions in regard to sustainability. It focuses on the development of an extended human identity and self-understanding, an integrated personality. The crucial aspect is that humans are relational beings, necessarily embedded in and constituted by the sustainability relations, and able and required to develop these relations. This requires the recognition, practice, and development of all potentials of the sustainability relations, as well as the development of the person in regard to her relational existence. The sustainable person would recognize herself as inevitably located within the sustainability relations and having the potential to develop competences and virtues in light of these relations. Her self-actualization would occur within and through these relations, constituting them and being constituted by them.

The above developed foundation of sustainability ethics enables us to recognize the ethical issues in the context of sustainability and to properly formulate ethical problems. The approach is primarily focusing on the role and morality of the person, and not on specific ethical rules or duties. However, the proper understanding of the relational aspect of sustainability ethics will help us to recognize and analyze all kinds of specific ethical issues surrounding the issue of sustainability. Particularly, it will help us to identify and address crucial ethical trade-offs, whose avoidance or minimization is one important ethical challenge in regard to sustainability. A simple example for this challenge is the case of biofuels produced by crops. If we consider

[14] There are two possible understandings of the sustainable person. First, the sustainable person can be understood as a real claim to our personal individual existence and being. Second, it can be understood as a theoretical claim, as a necessary precondition: Only if we conceptualize the human being as a sustainable person are we able to deduce all requirements of sustainability given by this definition. All other conceptions of the human being and its behavior, such as those given by economics, biology, etc., are not able to do so (see Becker 2006, which provides a discussion of this theoretical interpretation). In the context of this book, I refer to both interpretations.

this case within the framework of the sustainability relations, the ethical trade-offs are evident. Biofuels from crops are a renewable energy source and an alternative to fossil fuels. Their production has recently been supported in many countries and continues to increase. Biofuels have been said to be advantageous for a continuous supply of fuels and also for the protection of the global climate. Both of these aspects could be evaluated as ethically good in regard to future generations. However, there are obviously serious ethical trade-offs in the context of the sustainability relations: If crops are used for biofuels, they cannot be used to feed people. The result is a heavy negative impact on the living conditions and the well-being particularly of poor people currently living on Earth. Furthermore, if produced in monocultures, biofuels may also be judged as negatively affecting nature. In other words, we can see that relations (i) and (iii)—the relation among contemporaries, and the human-nature relation—are heavily affected in a negative way. Thus, biofuels can not be said to be a sustainable solution, although it is a renewable—i.e., continuable—energy source. This example also makes evident that sustainability does not just mean continuance, but continuance that supports the integrative development of the three sustainability relations.[15]

References

Aristotle. (2000). *Nicomachean ethics* (R. Crisp, Trans. & Ed.). Cambridge: Cambridge University Press.

Becker, C. (2003). *Ökonomie und Natur in der Romantik. Das Denken von Novalis, Wordsworth und Thoreau als Grundlegung der Ökologischen Ökonomik*. Marburg: Metropolis.

Becker, C. (2006). The human actor in ecological economics: Philosophical approach and research perspectives. *Ecological Economics, 60*, 17–23.

Becker, C. (2008). Thoreau's economic philosophy. *The European Journal of the History of Economic Thought, 15*, 211–246.

Becker, C. (2011): *Ethical aspects of the energy issue* (SSRN Discussion Paper). Available at: http://ssrn.com/abstract=1768194

Becker, C., & Manstetten, R. (2004). Nature as a you. Novalis' philosophical thought and the modern ecological crisis. *Environmental Values, 13*, 101–118.

Becker, C., Faber, M., Hertel, K., & Manstetten, R. (2005). Malthus vs Wordsworth: Perspectives on humankind, nature and economy. A contribution to the history and foundations of ecological economics. *Ecological Economics, 53*, 299–310.

Callicott, J. B. (1989). *In defence of the land ethics. Essays in environmental philosophy*. Albany: State University of New York Press.

Emerson, R. W. [1862](1998). Thoreau. In N. Baym et al. (Eds.), *The Norton anthology of American literature* (3rd ed., pp. 1202–1215). New York/London: Norton.

Hume, D. [1772](1998). *An enquiry concerning the principles of morals*. Oxford: Oxford University Press.

Leopold, A. [1949](1966). *A sand country Almanac*. Oxford: Oxford University Press.

[15]See Becker (2011) for a detailed analysis of the ethical aspects of the energy issue, and particularly of biofuels.

McIntosh, J. (1974). *Thoreau as romantic naturalist. His shifting stance toward nature*. London: Cornell University Press.
Nash, R. (2001). *Wilderness and the American mind*. New Haven/London: Yale University Press.
Rese, F. (2003). *Praxis und Logos bei Aristoteles. Handlung, Vernunft und Rede in Nikomachischer Ethik, Rhetorik und Politik*. Tübingen: Mohr.
Rousseau, J. [1755](1995). Discours sur l'origine et les fondements de l'inégalité parmi les homes. In J.-J. Rousseau, *Schriften zur Kulturkritik*. Hamburg: Meiner.
Shaw, B. (1997). A virtue ethics approach to Aldo Leopold's land ethics. *Environmental Ethics, 19*, 53–67.
Smith, A. [1759](2000). *The theory of moral sentiments*. Amherst/New York: Prometheus.
Thoreau, H. D. (1992). *Walden and resistance to civil government*. New York/London: Norton.
Thoreau, H. D. (2002). *Wild apples and other natural history essays*. Athens/London: The University of Georgia Press.
Wordsworth, W. [1814](1936). *The excursion; the recluse* (Part I, Book I). Oxford: Clarendon Press.

Chapter 10
The Structural Dimension of Sustainability Ethics

The sustainability relations are not constituted by individual personality, competencies, and practices alone. They are also constituted by social structures and mechanisms, as I have discussed in detail in Part II. The sustainability relations are, to an important extent, indirect and mediated relations, and to establish them in a stable and encompassing way requires certain structural means and mechanisms. The relational dimension of sustainability ethics cannot just refer to the individual and her ethical role within the sustainability relations, but must be complemented by a *structural dimension* of sustainability ethics, which addresses the ethical role of social systems in regard to the sustainability relations. The basic challenge of this structural dimension is to analyze what are good social structures in the context of sustainability, and what are not.

The basic ethical criteria for what are good structures result from the considerations about the sustainable person and the excellence of the sustainability relations (Chap. 9). In general, social structures have to be compatible with the relational dimension of sustainability ethics, with the cultivation of sustainable relationships. That means that they would allow the full development of the sustainable person and of the sustainability relations. They should support and foster the development of individual morality in regard to sustainability and not suppress or impede it. We need social structures and mechanisms that support the development of the sustainable person and of excellent sustainability relations. We should avoid all structural designs that endanger the development of the sustainable person, i.e., the development of the full potential of human being and the sustainability relations.

In fact, today all sustainability relations are to a large extent constituted and governed by the currently existing meta-structures (see Part II). In this chapter, I first provide an ethical critique of the existing meta-structures. I will argue that they are not fully compatible with the relational ethics of sustainability. The main reason for this is that they show an inherent tendency to reduce the human being and her relationships in certain ways and entail basic evaluations that are in conflict with a relational-based sustainability ethics and, particularly, the concept of the sustainable person. Second, I will argue that we need a careful analysis and redevelopment of the existing meta-structures rather than a revolution of them. We need to redevelop

C.U. Becker, *Sustainability Ethics and Sustainability Research*,
DOI 10.1007/978-94-007-2285-9_10, © Springer Science+Business Media B.V. 2012

the meta-structures with an eye toward the sustainability relations and the individual's potential to develop as a sustainable person with and through them. For this, I will provide some guidelines.

10.1 An Ethical Critique of the Existing Web of Meta-structures

The very meaning of the modern concept of sustainability implies that the ethical issue of sustainability is a matter of the development of excellent sustainability relations and an appropriate individual identity as a sustainable person. At the same time, this requires adequate social structures to implement and support excellent sustainability relations. The crucial question at this point is whether the currently existing global structures, which I have described as meta-structures, are able to do so.

The meta-structures work like filters through which we are related to ourselves, other humans, and nature. By their specific design, i.e., by their basic assumptions and evaluations, their driving forces, and their institutionalization, they heavily influence the way in which the sustainability relations are established and realized (see Chap. 8). By this, the meta-structures constitute an important condition for the individual's development and actualization within the sustainability relations and for the development of the relations themselves. I will argue that the currently existing meta-structures have a negative impact in this respect. They cause a problematic reduction of both the individual person and the sustainability relations. This simultaneously impedes the full development and actualization of the sustainable person, and the development of excellent sustainability relations. The task in this chapter is to provide a detailed ethical analysis and critique of the currently existing meta-structures. That means identifying what specific characteristics of the meta-structures negatively affect the sustainable person or the sustainability relations, and in what ways. The precise identification of the ethically problematic structural elements is a prerequisite to discussing options for a redesign of the current meta-structures.

10.1.1 The Reduction of the Human–Nature Relation

Against the background of the ethical guidelines developed in Chap. 9, it becomes obvious that the existing web of meta-structures has an ethically problematic impact on the human–nature relation. The meta-structures impede the development of a sustainable person and of an excellent human–nature relation. This is caused mainly by three problematic characteristics of the meta-structures: (1) the basic assumptions about the human being and nature made in all three meta-structures, (2) the basic evaluations underlying science and technology (i.e., the Baconian worldview), and (3) the dynamics of increasing complexity and influence of all three meta-structures.

1. In all three meta-structures the human–nature relation appears as a relationship between the human being as a rational being and nature as an object of humans' rational recognition, determination, control, and use. In modern science, nature is understood as an object of human recognition. Within this framework, science is striving for the determination of nature by human rationality insofar as it refers to causality and laws of nature (Sect. 7.1). In the sphere of technology, nature is an object of management that humans are enabled by their rationality to control. Science and technology together constitute the theoretical and practical means of recognition, determination, and control of nature. Finally, in the sphere of the economy, nature occurs as an object of economic rationality. It is a means for satisfying human needs. It is a good or production factor that enters the utility or production function and becomes an object of rational utility or gain maximization.[1]

   This perspective excludes other ways of relating to nature, such as by emotionality, creativity, and personal encounter (see Sect. 9.1). It also disregards other forms of rationality, such as practical wisdom and reason. This is in contrast to the basic human understanding underlying the sustainable person (see Chap. 9). The scientific approach leads to an abstract, artificial, and impersonal relation with nature, guided by its theoretical frameworks and methods. Natural sciences are not characterized by approaching nature with openness for its otherness, individuality, or own self. They approach nature to explain, systematize and determine it by rational means. There is indeed a tendency to more and more theoretical and abstract explanations and systems within science. For instance, in biology we have seen a huge shift from traditional research areas, such as zoology, to genetics and biochemistry. The scientific perspective on nature disregards personal, subjective experience and knowledge of nature. This impedes development of attentiveness and receptiveness for the other self in nature and, with this, the development of environmental virtues and a personal identity of interdependence in regard to nature. The whole scientific-technological perspective is based on the idea of autonomy of the human being—of its *independence* from nature. This is, of course, in contradiction to the characteristics of a sustainable person, for which *interdependency* is a constitutive element of its personal identity (see Chap. 9).

   All three meta-structures tend to interpret and realize the human–nature relation as a relationship between nature as an object of human thought and activities, approached primarily by scientific or economic rationality. The relationship is actualized merely on the rational level. The human being is conceptualized as a being characterized by certain types of rationality, striving for independence from nature, striving to overcome natural conditionality of human existence. This is in contrast to the sustainable person—to its identity as an interdependent

[1] For a detailed discussion of the representation of nature within economic thought, see Becker (2007) and Baumgärtner et al. (2006).

and natural contingent being, its attentiveness and receptiveness for nature as another self, its environmental values, such as respect for nature, and its underlying understanding of the human being as a rational, emotional, and creative being. The individual who is acting and thinking to a large extent within the context of the meta-structures as they currently stand is indeed in danger of misconceiving and missing the potential of her being, of missing the possibility to develop an identity as a sustainable person and of developing the sustainability relations in an excellent way.

2. The basic evaluation underlying science and technology is that the rational approach to nature and the striving for its domination and control—i.e., the striving for human independence and unconditionality—is good. This means that these meta-structures indeed establish an *implicit structural ethics* that is in contradiction to the relational dimension of sustainability ethics. From the standpoint of relational ethics of sustainability developed in Chap. 9, it is indeed not good to strive for the self-identity and the human–nature relation promoted by modern science and technology. This means that we have to find a solution for getting the current structural setting of modern science and technology in congruence with the requirements of the relational sustainability ethics.
3. The inherent dynamics of the meta-structures lead to an increase of their complexity and to an increasing influence of them on the individual and her sustainability relations. The increasing complexity and the increasing influence are ethically problematic because both increasingly impede the possibility of individuals to develop an excellent relationship with nature and a respective personal identity.

   The increase of complexity in science, technology, and the economy results in an increase in artificial and indirect relationships with nature. An example is farming and food production. The increase of technological and economic complexity in the field of food production leads to an alienation of the consumer and even the farmers from nature. For example, a highly specialized European chicken farmer buys the chicks from another farmer, specialized in the breeding of chicks. He feeds them with soya beans stemming from a region where he might never have been, e.g., Brazil, and has never seen how they are produced there. He holds his chicken a few weeks in a hall with artificial light, air-conditioning, and feeding machines before selling them to a company for slaughtering. Possibly he has never acknowledged the chickens individually during that time. The consumer eating a chicken might never have seen one alive, nor does he know how this one has been "produced." It is indeed a *production* of food, in which nature is degraded to a mere economic object and humans are reduced to mere rational utility maximizers. With this, the human–nature relationship is degraded and disturbed as a result of the technological and economic system governing it.

   Moreover, the dynamics of the meta-structures result in their increasing influence on societies and on individual lives. This continuously decreases the possibility for the individual to develop an excellent relationship with nature and a respective personal identity *beyond* the influence of the meta-structures.

   That is to say that in a world of increasing scientific, technological, and economic mechanisms and structures, with increasing degradation of nature to an object within the meta-structures, with increasing distance to and artificial encounters with nature, the individual potential to develop a relationship with, and personal identity and attitude toward nature, is steadily decreasing.

10.1.2 The Reduction of the Relationship with Future Generations

The current meta-structures also have an ethically problematic impact on our relationship with future generations. This mainly results from the impact the meta-structures have on the transfer of knowledge and artifacts—i.e., on the temporal cultural stability—as well as on the transfer of natural conditions. With this, they impede the development of an individual's identity as a temporal, culturally contingent, and interdependent being, and they impede the development of an excellent relationship with future generations. Moreover, the effects of the meta-structures also impede the very possibility of future generations to develop excellent sustainability relations and identities as sustainable persons.

The transfer of knowledge to future generations, which is a crucial means for realizing the relationship with future generations, today is heavily affected and governed by modern science. Science creates knowledge, defines the field of knowledge, and, with this, affects what is taught at universities and schools. I have argued that scientific research in fact results in an ongoing, coupled increase of knowledge and ignorance. This is combined with an increasing complexity of knowledge and a loss of stability of knowledge—a decreasing half-life of knowledge (see Sect. 7.1). These effects are mainly the result of the problematic epistemological basic assumptions underlying modern science, which lead, together with the basic evaluations and the institutionalization of modern science, to a dynamic of increase of both scientific knowledge and ignorance. However, the coupled increase of knowledge and ignorance, and the decreasing half-life of knowledge, impede the transfer of knowledge over time. The effective transfer of knowledge requires a certain stability and limitation of knowledge. It is neither possible nor fruitful to transfer a permanently increasing amount of knowledge to the future. The transfer of knowledge requires individual persons who are able to develop their potential to understand and actualize the knowledge, and it also requires the development of means for storing and transferring the knowledge. Both the means of transfer and the ability of adoption are limited. This is the reason why we face an increasing loss of previously acquired knowledge. Thus, the knowledge transfer to future generations becomes unstable, and, with this, so does the relationship with future generations.[2] The main

[2] This is also a technological problem. Knowledge has to be stored on media. The huge amount of current knowledge requires digital media, which have a rather short half-life time. Moreover, there is certain technology and know-how required to read those media. The complexity and rapid change in the field of technology makes it very difficult to guarantee that this will still be possible in the more distant future (see also The Long Now movement; www.longnow.org).

cause for these problems is the still dominant assumption that there is a limited amount of ignorance given in the world, which we can (and should) reduce by scientific research. This leads to the assumption that the guideline for research is externally and objectively given by unambiguously given ignorance which must be addressed and reduced. By this, it is not recognized that we, as culturally interdependent and responsible beings, are claimed (called upon) to give that guideline by ourselves. We must guide the process of knowledge generation as an indefinite and open creative process by ethical guidelines and in regard to our relationship with future generations. This is one of the requirements of a sustainable person, which is impeded by the epistemological assumptions of modern science. Established modern science, as it is currently designed, endangers the stability of cultural identity in the long run, and with this, a pillar of sustainability.

However, it is the combination of science and technology as designed by the Baconian worldview that causes more serious ethical problems. The problematic impact of both on the relationship with future generations results not just from the long-term risks of some technologies, such as nuclear technologies or genetic engineering. Of course, such technologies or their remaining parts may endanger other generations at some point in the future. But there is a more hidden danger, which results from the interrelation of technology and science, and their coupled dynamics. The increase of knowledge is coupled with an increase of application of that knowledge in the field of technology. For instance, we do not merely create theoretical knowledge in genetics or chemistry, but use that knowledge to develop technological applications affecting the real world. However, there is the problem of knowledge transfer to future generations described above. This could potentially lead to a situation in which many technological developments and their consequences last into the farther future, but the respective knowledge to understand them will be lost. Thus, future generations may be confronted with our technological inventions without receiving the related knowledge. This will reduce their ability to understand the results of today's technological impacts and residues, and to deal with them. That effect is compounded by the increasing complexity of the technological applications. This increase makes it impossible to oversee the impacts on future generations, particularly the longer term environmental effects of technological developments. For instance, the huge amount of chemical agents or genetic modifications we have created so far may lead to a lot of unforeseen effects and side effects in the environment in the long run, and thus, to impacts on future generations we cannot imagine today.

The ethical problem here is that we cannot develop a serious and clear relationship with future generations on the structural basis of modern science and technology. We cannot develop an attitude of responsibility and care toward future generations if we are not able to foresee the results of our action, and we cannot guarantee that we can transfer our technology and knowledge to them in a way that will allow them to deal with it. This is the result of the very design of modern science and technology, which suggest a control and determination of the future. It leads to a *misconceived temporal self-identity* of the currently living generation, which is based on the idea of rational control and determination of the future and future generations. Indeed, the current prevalent reaction to the increasing problems of sustainability is to remain within the Baconian worldview and strive to control technology and its effects through more research and further technologies. This impedes the development of an identity as a sustainable person, as a temporally interdependent and responsible being. As long as we remain in this scientific-technological worldview, we cannot develop an excellent attitude of responsibility, attentiveness, and care toward future generations.

A similar problem occurs if we consider the impact of the current economic system on our relationship with future generations. In the field of the economy we act as economic persons, and the whole economic system is based on self-interest and focused on short-term results. This also impedes the development of a temporal identity and an attitude of responsibility and care for future generations. The economic person has no systematic interest in future generations, and the latter cannot participate and bring their interests into the current market system. In fact, the modern economy results in an excessive use of resources, production of waste, and transformation of nature. If sustainability in regard to future generations is discussed today in economic terms, one remains in the general framework of the economic view. One discusses what economic entities and possibilities shall be provided to future generations. Economics, e.g., analyzes the intertemporal maximization of welfare, or the conditions of nondecrease of welfare or capital over time. One prominent economic discussion on sustainability concerns capital transfer over time: can and should we substitute natural capital by human-made capital? Are we allowed to use up non-renewable resources if we replace them with human-made capital, or should we maintain each kind of capital separately for future generations? The first alternative is called *weak sustainability*, the latter *strong sustainability*.[3]

Within current scientific-technological and economic frameworks, our relationship with future generations is understood and actualized in terms of rational control and transfer of economic entities and welfare. This is a reduction of that relationship and of the personal identity of the sustainable person. The relationship with future generations is not just about rational control and planning of the future, or about economic aspects. It is about attentiveness, respect, and care for individual persons in the future—their otherness and particularity, their possibilities for their own unique development. This is particularly evident in the case of our children, where neither rational control nor the bequeathing of goods is essential for an excellent relationship with them, but personal respect, attentiveness, and love, care for their own and free development, etc., take precedence.

Finally, a remark on the well known and often discussed problem of appropriate transfer of natural conditions to future generations. It is often argued that we should sustain nature or certain environmental services for future generations. The problem,

[3] For an overview of economic approaches to sustainability, see Berg and Hofkes (1999). For the discussion on weak and strong sustainability, see also Neumeyer (2004) and Ayres et al. (1998). For ethical aspects, see particularly Ott and Döring (2008).

however, is what exactly should be sustained, and for what reasons. A common argument refers to the survival of humanity and makes it the crucial norm for what we should do in regard to sustainability. However, this is a rather reduced ethics, because the crucial ethical question is not about survival but rather about a good life. This means that in regard to future generations we should ask if the way we have developed our relationship with them, and particularly the conditions and social structures we provide them, will allow them to live a good life. In terms of sustainability ethics, that particularly means if they will be able to develop a sustainable identity and excellent sustainability relations by themselves. It is at least doubtful that they will be able to do so in the natural as well as cultural environments we have created for them.

10.1.3 The Reduction of the Relationship Between Contemporaries

In regard to the relationship between contemporaries, the most important and problematic meta-structure is the economy, at least as it is organized and designed today. The economy today plays an important role in individual life and self-identity, as well as in the relationships between contemporaries on the societal and global level. By its very design, the modern economy impedes not only our relationships to future generations and nature, but also heavily affects the relationship between contemporaries. This is mainly caused by the basic model of the human being as an economic person, the basic assumption about the systematic coordination of social interactions, the driving forces of nonsatiation and economic growth, and the underlying evaluations that both self-interest and economic growth are good, i.e., that they will always lead to an improvement of human well-being when actualized in the context of a market economy. I want to discuss these aspects by referring to both (1) the economy as a meta-structure given in reality, and (2) economics and its theoretical underpinning.

1. The individual person and her everyday life, thoughts, and actions are deeply embedded in the economy. This affects individual self-identity, and persons increasingly understand themselves as *economic persons*, i.e., as self-related rational utility maximizers. This generally may not occur as an explicit definition of one's own identity, but may actualize itself in different ways. One may regard consumption of goods as essential for one's own well-being and correlate a better life and happiness with increasing consumption. One may define personal development by economic success and personal income. Or, one just may be fully occupied with business activities and economic necessities. With this, there is also a tendency to regard others as mere economic persons and relate with them merely on an economic level, i.e., based on economic rationality and self-interest. The interaction with others may even be seen as a kind of strategic game. Overall, economic principles, rules, mechanisms, and structures, such as the market, become the dominant guiding principles of social interaction and interrelations.

However, understanding oneself and others merely as economic persons, and relating to others on this basis, impedes the development of an identity as a sustainable person as well as the development of excellent relationships between contemporaries. It rules out other dimensions of human existence and interaction, such as understanding the human being as a social, ethical, creative and rational being in a broader sense. It restricts attentiveness and receptiveness for others to the issue of recognizing their self-interests and does not allow for a systematic and encompassing attitude of responsibility or care for others. Thus, if personal self-identity is merely or to a large extent defined by being an economic person, the sustainable person and her relationship with contemporaries are endangered.

Additionally, we must take into account the basic evaluations underlying the modern economy. These are the evaluations that self-interest is good and that non-satiation and economic growth are good. In the context of the market economy all this will, in principle, improve human well-being and society. These evaluations are also the main driving forces of the economy and cause its dynamics. This results in an ongoing increase of the influence and the complexity of the economic system. With this, however, there also is an increasing negative impact on the potential for personal development as a sustainable person, as well as on the relations between contemporaries. Self-identity as economic persons and economically dominated relationships are not mere static issues. With the increasing influence of the economic sphere, individuals are increasingly involved in economic issues. It becomes more and more difficult to elude the economic sphere, which becomes increasingly influential and important for the individual and her relationships. We can recognize the increasing influence also on the societal and political level where social and political issues, such as the design of the welfare state or the educational system, are discussed to an increasing extent in economic terms (Becker and Jöst 2007).

A particular ethical problem results from the increasing complexity of the economic system on the global level. The increasing complexity of the global economy generates global relationships that are more and more indirect, mediated by economic mechanisms, and that are more and more difficult to recognize for individuals. By our individual economic activities we are in fact related in a complex way to many people in the world without knowing exactly how. This thwarts a deliberate establishment of personal ethical attitudes in regard to global relations. We cannot develop attentiveness, care, or responsibility in regard to relationships that we are not able to recognize and understand. For instance, if we want to consume in a responsible manner and to take care of the environmental and social conditions by which goods have been produced, we are faced with the serious problem that we are not able to get all the necessary information because of the complexity of global production processes.

In fact, we all are interrelated on a global level in a complex way by economic processes and structures, of most of which we are largely unaware. Business has indeed become the most important means by which we are related to other persons on a global level. This results in a situation in which our global relationships are mainly *tacit* and abstract ones and to a decreasing extent concrete encounters, dialogues, or communication. Current global relationships are largely reduced relationships between economic persons and do not allow for recognizing and respecting the specific otherness and cultural identity of people. We are economically related without being able to fully recognize and communicate cultural identities and differences. The economic person is thought to be the same all over the world—all are self-interested rational utility maximizers—and it is assumed that economic decision making and mechanisms are the same in any place. The economic perspective on global relationships focuses on the sameness of all actors and not on systematic differences such as cultural differences.[4] The self-identity of an economic person erases cultural identity and intercultural competence. It impedes the development of an identity as a sustainable person, as a culturally interdependent and contingent being, and it impairs the ability to relate to others on this basis.

Global relationships today are actually governed by a few reduced means. The most important of these is the economy. Other specific means, which are partly influenced by the economy, are mass tourism, mass media, war, and at the margin—if we consider the number of people involved—political contacts. Real and encompassing intercultural relationships and dialogues cannot be established by these means. We hardly have any dialogues or genuine intercultural communication at all. We have more tacit relations, governed by complex economic interdependencies, largely one-sided and unidirectional mass media contacts, non-communicative mass tourism, and war. Global relationships are indeed disturbed, if not broken. This is not just an issue in the context of sustainability, but may be one of the most important causes of current international and intercultural conflicts. However, the recent increase in worldwide Internet communication may have a potential to overcome the dominating one-sided and reduced means, which have governed global relationships so far, and may potentially lead to a new way of global interrelationships more in line with the excellence of sustainability relations and the sustainable person.

2. It is important to also discuss the perspective and role of economics. Economics is crucial for the theoretical underpinning of the modern meta-structure economy, and strongly influences the understanding and practice of business. Moreover, economics plays a crucial role in the academic study of sustainability as well as political decisions. For instance, the politics of Thatcher and Reagan were much influenced by economic advices based on neoclassical economics.

   One of the most important achievements of modern economics has been its recognition and theoretical analysis of the meaning of structural mechanisms for social interactions, particularly the recognition of the difference between individual intentions and overall social results of actions (see Sect. 7.3). With this,

[4] However, if we consider the foundations of the modern economic worldview as a particular result of European thought, we may regard the modern economic perspective as a kind of cultural imperialism.

economics developed a specific and very influential theoretical approach to social interactions and interrelations. Most prominent in this context are invisible-hand theories (see Sect. 7.3), which explain social interactions and their results not as matters of individual intentions, but rather as deeply influenced and even designed by social structures. Certain settings systematically lead to certain results without being explicitly intended by any of the individuals acting within them. The most prominent setting is the market, in which everyone is following his or her own interests and, by the very design of the market, the overall welfare is systematically improved, even if this was not the intention of any of the individuals acting within the market system (see Sect. 7.3).

Although most current economists would deny that economics explicitly deals with ethical issues, economic theory without doubt has ethical relevance. It entails at least an ethics of laissez-faire, in which the individual is said to be free to follow his or her self-interests and individual preferences and need not develop a systematic individual responsibility or care for the social consequences. As long as the design of the social structures is appropriate, it is good to concentrate on one's own self-interests, at least within the sphere of the economy. From this point of view, the ethical issue is not primarily a matter of individual morality, but of the proper design of social structures. Originally, this view was restricted to the sphere of the economy (Smith [1776]2000). However, because modern economics tends to be a universal theory about individual decision making and interactions, the invisible-hand theory and related ethical implications became in principle relevant for all social interactions. The invisible-hand explanation has been extended, and the orientation of social interactions is generally seen as an issue of designing social structures, not as an issue of individual morality. In this view, the design of social structures becomes crucial for ethics, and not the individual's moral considerations and attitudes. Eventually, this results in the idea of a mere structural ethics, which solely focus on the design of economic institutions (see, e.g., Homann 2002).

However, this perspective is ethically problematic, particularly in regard to sustainability ethics. Modern economics entails a tacit structural ethics that is not compatible with the relational dimension of sustainability ethics. One crucial problem is the underlying view that human beings are mainly motivated and guided by self-interest. Individuals are assumed to be merely self-interested and to have arbitrary preferences, and individual freedom is understood as the freedom to follow these preferences.[5] In the economic system individuals interact voluntarily and agree to transactions, contracts, etc., only if it is in their self-interest. In a voluntary economic transaction all participating actors are better off (or at least not worse off) than before. Otherwise they would not have agreed to the transaction. In this specific way the economic system is said to be a sphere of individual freedom.

[5] The assumption about the independence of individual preferences—combined with the specific concept of economic rationality—is indeed in contrast with the project of most traditional moral philosophy, i.e., the project of finding interpersonal agreement on fundamental values by rational consideration. In this respect, the economic perspective tends towards an ethical relativism.

This is a rather abstract and simplified conception of the economy. It models a simple interaction between actors who have equal access, power, and information. It does not take into account that there are patterns of domination, asymmetrical power, and even oppression, particularly in the sphere of the global economy. Moreover, this economic model is not appropriate for approaching relationships in which the other part cannot participate in interactions, such as our relationship with future generations or nature. The underlying understanding of the human being as a self-interested rational utility maximizer is a specific model of the human being, which may be appropriate for analyzing some specific economic settings. If extended to a universal concept of the human being or used to approach the issue of sustainability, the *homo economicus* is misleading and counterproductive (see also Becker 2006).

A second crucial ethical problem of the economic perspective is the underlying pattern of control. Like science and technology, economics leads to, and proceeds based upon, an idea of control and determination. From the perspective of economic theory humans are by their very nature self-interested, and the market system is a perfect structure to coordinate their interactions. Most social and economic problems are interpreted as market failures, and the solution is seen in market instruments, which reestablish the market mechanism and its efficiency. This leads to the idea of a systemic control of social interactions through the proper design of markets and other frameworks of social interaction. This perspective stresses the importance of the systemic level for proper social interaction and regards individual morality as, at best, a secondary feature, or even as not relevant at all. One just needs to properly design the system in which the individuals act and define the right rules or incentives, and this is enough to guarantee the right actions and social results. There is no need for individual morality, responsibility, or practical wisdom. The only requirement of the individual and her morality is that she follows the rules or incentives.

However, this idea of systemic control of social actions and interactions is neither ethically desirable nor practically possible. A merely systemic ethics would indeed be a dangerous illusion. The exclusive use of rules and economic incentives for guiding individual behavior—e.g., guide individuals through eco-taxes to behave in environmentally friendly ways—would not work as a universal ethical approach. First, current systems such as the global economy are too complex and too dynamic for a mere systemic control. There is no perfect framework or set of rules that could properly govern interactions on a global level. Second, if people would actually act as mere *homines economici* and refer merely to their self-interests, they would always find ways to diverge from the given systemic setting. In the complex and dynamic economic system, they would always find loopholes in the given rules, which they could use to their own advantage. Each case of divergence would require an improvement and extension of the framework and the rules.

However, the more extended and complex the framework of rules, the more options to diverge arise. An example would be the German or US tax system. The system is designed in increasingly sophisticated ways to prevent tax evasion.

However, the result is that the more complex the rules, the more ways are found to circumvent them. The conclusion from this observation is not that humans are by nature self-interested and that we should strive for a better system to control and guide them, but rather that we need an individual ethics as an important complement to a systemic ethics. The economic perspective does disregard this conclusion, and moreover impedes the development of a conception of individual morality, e.g., such as the sustainable person, by its underlying concept of the human being. The *homo economicus* is not an appropriate basis for the development of a theory of ethical decision making. The *homo economicus* is arbitrary to values or norms. Its rationality does not allow for any objective value judgments.

Thus, what we need in regard to sustainability and sustainability ethics is not an *extension* of the economic approach to these issues, but rather a *redesign and reorientation* of it. The market and other mechanisms are without a doubt very important social coordination mechanisms, and they have been to some extent very successful. However, to use them in a constructive way for creating sustainable development requires adequately transforming them and embedding them into a broader ethical framework. This requires a modification of the meta-structure economy and the underlying assumptions of economics in such a way that they become compatible with sustainable persons and excellent sustainability relations.

10.1.4 Conclusion

The existing meta-structures set us, in specific ways, in human relations, particularly in the sustainability relations—our relations to our contemporaries, nature, and future generations. As I have argued above, this results in a reduction of these relationships, as well as in a reduction of the individual moral potential in regard to these relationships. Constrained and governed by the current meta-structures, the human being is a *relationally disturbed* person in a fundamental sense: she is not able to fully recognize and properly actualize the sustainability relations, and to understand, develop, and actualize herself as a sustainable person within.

In addition, we must take into account that the meta-structures show a dynamics of increase driven by their driving forces. The meta-structures, therefore, increasingly influence and even dominate individual action and thought, as well as the sustainability relations. The meta-structures have an increasing impact on the ethical aspects of these relations and, thus, on the ethics of sustainability. This makes it even more urgent to redesign them in a way that would make them more compatible with the relational dimension of sustainability ethics.

It is important to recognize that the ethical critique given above holds to some extent even if one does not refer to the relational ethics I have developed in Chap. 9, but rather to traditional moral theory. This is particularly true for the critique of the increasing complexity of the meta-structures. Such increasing complexity reduces the possibility to foresee the fuller consequences of our actions, to predict who will be affected by what action, and in which ways. This is ethically problematic against the background of all consequentialist ethical theories, i.e., all ethical theories, particularly utilitarianism, that refer to the consequences of actions as crucial for ethical analysis. Increasing complexity renders the consequences of a given action increasingly dubious and unclear. This undermines individual morality and ethical guidance on the basis of consequential ethics.

Another example for the ethical problems of the meta-structures even in regard to traditional ethics is the concept of responsibility. This concept requires that we know at least to some extent the results of our actions, in which ways we affect whom, and how we are related to whom. We must be able to act intentionally in a sensible way. However, if we are faced with a world in which this is not sufficiently possible, in which complexity and unforeseen side effects of actions dominate, the idea of individual responsibility is undermined (see Petersen and Faber 2006).

Morality in regard to sustainability, thus, cannot solely be founded in individual morality. It needs adequate social structures that would support it, and that, in turn, it would support. Although the meta-structures are historically developed and given, they are not a given *fate*, but can be further developed and changed. In this respect we may say that we have a *meta-responsibility* to develop the social and global structures for our actions and relations in ethically appropriate ways, such that these structures foster individual morality and do not impede it. It is, therefore, not fruitful to have an *end-of-the-pipe ethics,* i.e., an applied ethics that simply reacts to issues that occur within the given structures, but does not critically analyze the structures themselves. For instance, a business ethics that merely accompanies the ongoing processes of business and takes the current conception of the economy as given would be insufficient (Becker 2011). We do not need a simple applied ethics in regard to the future design of a sustainable society. We rather need a sustainability ethics that analyses the very structures, the very preconditions and settings of the social framework, in light of relational ethical principles. We need a systemic ethics that asks what would be a good science, a good technology, a good economy; that is, an ethics that provides ethical guidelines for an appropriate redesign of meta-structures in regard to compatibility and integration of the structural and individual ethical dimension of sustainability.

10.2 Guidelines for a Redesign of the Meta-structures

It is more appropriate to redesign the current meta-structures—in the sense of developing and improving them over time—than to strive to completely erase them and to invent entirely new social structures. This is for three reasons. First, the critical analysis demonstrated that not *all* aspects of the meta-structures are ethically problematic, but only *specific* characteristics of them. The meta-structures are well-established structures that also possess sophisticated elements and mechanisms that may help to constitute excellent sustainability relations. For instance, the market has been proven to be a well-working mechanism for the coordination of economic interactions. It may be possible to keep this mechanism and redesign some of its problematic underlying assumptions and evaluations. The market is, for instance, not necessarily linked to the growth paradigm that is integral to the basic evaluations of the modern economy. Second, the approach of redesign of the meta-structures is in accordance with the relational identity of the sustainable person, particularly with her temporality and cultural contingency. The redesign approach accepts that the current meta-structures are to some extent part of the cultural and individual identity of currently living generations and that it is more in line with a stable and sustainable development to redesign and further develop the meta-structures than to try to invent and implement entirely new structures. Third, the approach of redesign of the meta-structures is the more pragmatic approach to the structural ethical problem in regard to sustainability. It is more realistic and practicable to develop and improve the existing, well established, and institutionalized structures than to implement entirely new ones. Nevertheless, we will have to change some of the very fundamentals of the existing meta-structures to get a sustainable future. This will by itself mean some kind of revolution. However, because of the difficulties that will inevitably already arise with the project of changing these fundamentals, it is of importance to try to keep as many elements and mechanisms of the current meta-structures as possible.

In the following, I will provide no detailed plan for a redesign, but rather some general guidelines. For this, I concentrate on three crucial ethical problems of the meta-structures, which have been identified in Sect. 10.1. First, the reduced understanding of the human being and of human rationality, which is—in different ways—the basis of all three meta-structures: The human being is conceived as a merely rational being, whereby rationality is defined in rather narrow ways as specific types of scientific or economic rationality. As a result, in the context of the meta-structures, all sustainability relations appear to be governed and controlled by these narrow types of rationality. The person conceived by these meta-structures is the separated, independent, and autonomous individual, rather than the relational and contingent person.

The second crucial ethical problem is the basic evaluations and inherent ethics of the meta-structures, which entail two core aspects: The ideal of individual independence and rational control of the sustainability relations, and the fundamental growth paradigm. The first basic evaluation suggests that the independence and autonomy of the isolated individual, as well as rational control of her sustainability relations, are ultimate aims; the second suggests that more knowledge as well as more goods is always better than less. Both will, according to these basic evaluations, supposedly lead to an improvement of human well-being.

The third crucial problem is the increasing influence and increasing complexity of all three meta-structures. This is caused by specific driving forces; in particular, the basic evaluations that more knowledge is better than less, and more goods are better than less, are driving the increase. But it is also the case that certain aspects of the institutionalization of the meta-structures foster that dynamic, or add their own inherent dynamic to it.

As a consequence, I suggest three main guiding principles for redesigning the meta-structures:

1. An encompassing concept of the human being and of human rationality
2. New basic evaluations
3. Stability and simplicity

1. The development of a sustainable person and excellent sustainability relations require as basis a broader understanding of the human being as a rational, emotional, creative, and communicative being and, with this, as a relational being. We have to make sure that the meta-structures support—or at least allow for—this understanding. For this, we need to implement a broader understanding of the human being in the basic assumptions of the meta-structures themselves, or, alternatively, embed the existing basic assumptions about the human being explicitly into a broader concept of the human being defined beyond the meta-structures. However, the most fruitful approach is to do two adjustments at the same time: Some redefinition of the basic assumptions of the meta-structures, which allows for making them compatible with an encompassing concept of the human being, and the embedment of these redefined assumptions into a broader concept of the human being.

   In other words and with the example of the economy: the identity as economic person should fit into the broader identity of the sustainable person. This requires some redefinition of the economic person and economic rationality. We need to reconsider our understanding of the economy and business. A starting point could be a reexamination of today's business beyond the abstract conception of modern economic theory. This would reveal that business is not merely about maximization of gain or utility, nor is it simply a matter of strategic games. Rather, business can also be understood as a process of creation of concrete products and competencies for society, as social endeavor of people working together, as management and organization of social structures and common work. Business and the economy, thus, are also about communication, creativity, and relationships (see Freeman 1984, 1994 and Becker 2008, 2009, 2011). If we understand the economy in this way, we may also redefine the economic person and her characteristics. It is then no longer the mere selfish rational utility maximizer, but the communicative and creative person who develops and organizes economic relationships and creates values within them. This requires certain personal abilities beyond modern narrow economic rationality. It requires the ability to develop personal character and economic relationships in an excellent way. It requires the development of specific business virtues, such as leadership qualities, professional integrity, and toughness in negotiations (Solomon 1992). One may say that this ultimately also requires the development of an identity as relational being and of basic virtues, such as attentiveness and receptiveness for others. Further, this understanding of business also requires practical wisdom to be able to develop and properly actualize these virtues. This view of business would lead to a broader understanding of the human being as basis for the economy (see also Becker 2011).

Aristotle already recognized the relational and related ethical aspect of the economy (Aristotle 1995; Becker 2009). For him the economy had an inherent ethical aspect, as it is concerned with the proper organization and management of specific economic relationships. He concluded that practical wisdom (*phronesis*), which he considered to be the relevant ethical type of rationality, is crucial for the economy as well. Practical wisdom allows the individual to organize and manage the economic relationships in a sensible way. For Aristotle, questions of what is sensible in regard to the economy, and what is the proper constitution and management of economic relationships, can be defined only with regard to the embedment of the economy into society. Practical wisdom has to understand and develop the specifics of economic relationships in the light of the general relationships of the whole society (Becker 2009). In modern terms: it would not be appropriate to develop the rules and virtues of business in separation from social morality. Of course, the virtues developed within the specific relationships of the economy can be specific *economic virtues*. However, these should never fully be isolated from one's overall personal identity or personal virtues. The economy is no neutral or separated interaction mechanism. As a sphere of common creation of goods, of development of economic relationships and personal identity, it is an integrated part of the overall development of relationships and personal identity within society.[6] Particularly, businesses, products, work, and aims cannot be justified merely by their function of maximizing profit or individual utility function, or their abstract contribution to overall gross domestic product. Rather, businesses, products, work, and aims have also to be justifiable by their concrete contribution to societies and the sustainability relations and their coherence with our self-identity as relational persons.

Understanding the economy and the economic person in the broader sense described above allows for an integration of the economic person and the sustainable person. The businesses, products, work, and aims of the economy become part of the individual's sustainable identity. It is a mutual process in which the development of economic identity and relationships is guided and takes part in the overall development of the sustainable person and excellent sustainability relations. This way, it is possible to redefine the understanding of the human being in the basic assumptions underlying the economy so that it is compatible with the concept of the sustainable person and feasible for the integration into the latter concept. It is ultimately the sustainable person who acts within the economy and should do so in the light of his or her sustainable identity.

Two remarks on the example of the economy: First, my suggestions result in an explicit inherent economic ethics, which is in accordance with sustainability ethics. I do not suggest an *additive* extension of the economic person and economic rationality, by which other forms of rationality and personal identity are

[6] See also Solomon (1992) and his virtue ethics approach to business ethics.

attached to the existing concept. I rather suggest an *integrative* extension, which reinterprets the economic person and economic rationality in a way that explicitly builds in an ethical dimension. This avoids a simple external ethical control of business and the economy—an *ex-post* or *end-of–the-pipe* business ethics—and inevitable conflicts between the implicit ethics of the existing economy and ex-post added ethical frameworks. Second, I hold that this is indeed not a merely theoretical concept, but that it is possible to change the understanding of business, the economy, and the economic personal identity in practice. This of course requires societal agreement and efforts in regard to several mechanisms. Education, laws, taxes, etc., can all foster the reconceptualization and re-understanding of the economy and the economic person. On all levels it has to be demonstrated that society and its institutions do not foster freedom for egoism, but rather the freedom to develop and actualize an identity as relational and sustainable persons. Society, and particularly politics, has the ability to change the very understanding of the economy and business that way. It is not the case that there are economic "laws" or requirements of the global economy that politics has to follow by necessity. Doing so is rather dangerous, as, e.g., the global financial crisis of 2008 has demonstrated. On the other hand, we have some successful examples of starting change. The recent rise of the topic of *corporate social responsibility* has already shown that there is an increasing societal demand to integrate business better and in ethical terms into society and had led at least to some changes in the self-identity of corporations and managers.

I have so far extensively discussed the redesign of the economy in regard to guideline (1) above. In science and technology, however, in principle the guideline also holds. We must reconsider the underlying understanding of the human being, and, particularly, its definition as being separated from nature. However, in this case the crucial problem is not a necessary conflict between the understanding of the scientific person and the sustainable person. It is, rather, the claim of totality of science and scientific rationality as the sole way of approaching the human–nature relationship. In principle, approaching nature by scientific rationality could be part of the identity of the sustainable person, if it is, from the very beginning, embedded into a broader concept of the human being and the human–nature relationship and guided by ethical orientation. However, this would require a redesign of the basic evaluations underlying science and technology, which impedes such a re-understanding of the scientific person and an inner ethical orientation of technology and science. The separation paradigm is strongly related to the evaluation that the ultimate aims of science and technology are independence from, and control and domination of, nature. To replace this with a paradigm of interrelation requires a new basic evaluation of science and technology, one which would allow for an inner ethical orientation of these structures, and a restriction of the claim of totality of scientific rationality as the sole approach to the human–nature relationship.

2. New basic evaluations for all three meta-structures are very important, and could be developed and implemented without completely changing these structures. One crucial aspect of all basic evaluations is the idea that the independence and

autonomy of the individual, as well as rational control of her relationships, are ultimate aims. In regard to the economy, I have already argued that this assumption can and should be replaced by a new understanding of the economic person, which entails an inherent ethical aspect and an economic identity as a creative, communicative, and relational being. In regard to science and technology, we have particularly to overcome the idea that their ultimate aim is human independence from and control of nature. We need a replacement of the independence and control/domination paradigm by a paradigm of interrelation and interdependence. We need a new understanding of science and technology as means for developing an excellent relationship between humans and nature. Heidegger (1977), for instance, suggested some kind of such a new understanding: A re-understanding of technology as an original expression of human creativity embedded in the more fundamental bringing-forth of nature. The latter is the inevitable condition and ultimate reference point of technology and science. We can reinterpret science and technology as creative processes of interaction with nature, with ultimate aims not of independence and control, but excellent relationship between humans and nature. That means that there is no longer a simple external orientation on the ideal of independence and control, but the orientation has to be developed in the context of the development of the sustainable person and excellent sustainability relations. Science, technology, and the economy have no ultimate aims or given orientations, but their orientation and development is an ongoing ethical challenge for the sustainable society.

The second crucial characteristic of the basic evaluations of all meta-structures is the underlying growth paradigm, which suggests that more knowledge as well as more goods is always better than less, and that an increase of both will lead to an improvement of human well-being. In the light of the critique of this paradigm given in Sect. 10.1, we should overcome this idea of automatism. We have to take into consideration the negative impacts of growth on the sustainability relations, as well as the problematic underlying assumptions about knowledge and ignorance, human preferences, and well-being. The growth paradigm—the old social contract underlying science, technology, and economics—needs to be replaced by a new paradigm of sustainability—a new social contract. More knowledge, technology, or goods can no longer said to be good per se. We rather need an explicit societal and ethical orientation for the production of knowledge, technology, and goods. This orientation results from the embedment of the meta-structures in the overall development of the sustainability relations and sustainable person. Again, there is no simple reevaluation criterion or rule, but the underlying basic evaluation of all meta-structures becomes part of the development of excellent sustainability relations and the sustainable person.

Problems of the growth paradigm and possible alternatives have already been discussed in regard to the economy, e.g., prominently by the Club of Rome or ecological economics (Georgescu-Roegen 1971; Meadows et al. 1972; Daly 1977, 1996; Becker et al. 2005). In contrast, there is much less discussion about the growth paradigm in regard to science and scientific knowledge, although it is likewise crucial for sustainability. In principle, neither the scientific ideas of recognition and knowledge generation, nor the economic ideas of rational agents and markets, require the growth paradigm by necessity. However, the growth paradigm has had a very strong influence on all structures and the actors within. Particularly, the specific institutionalizations of the meta-structures have supported the growth paradigm and the dynamic of increase and real growth of all meta-structures, which is difficult to overcome and change.

3. The third guideline addresses the dynamics of the meta-structures. It is important to avoid a problematic increase of their influence and complexity. We have to balance the dynamics of the meta-structures and to change those driving forces, which cause increase and increasing complexity, in orientation to a paradigm of stability and simplicity. This does not mean that there should not be a dynamic development and growth of these structures at all. It rather means that we have to avoid the current *automatic, omnidirectional, and uncontrolled* dynamics of increase and to replace it by an *explicit, deliberate, and oriented* dynamics. The change of the basic assumption about the human being within the meta-structures, and the change of the basic evaluation according to the guidelines discussed above, will already support this aim to a large extent. They will allow for an orientation of the dynamics by the identity and responsibility of the relational person.

   To change the dynamics of the meta-structures, however, we must also reconsider and redesign the institutionalizations of the meta-structures, which, as I have discussed previously, play a crucial role for their dynamics. There are a variety of institutional characteristics that result in internal incentives and individual attitudes fostering the automatic growth of the structures and their complexity. For instance, in science, internal evaluation processes give incentives for producing more and more papers; or, in the economy, a misinterpreted shareholder-value based evaluation mechanism of corporations and managers gives strong incentives for a specific dynamics of growth. To follow guideline (3), and implementing simplicity and stability, thus, requires a careful and detailed redesign of the institutional settings of the meta-structures. In regard to science, I will discuss this in more detail in Part IV.

The ethical critique and the guidelines for redesign of meta-structures are an analytical and theoretical approach to the structural dimension of sustainability ethics. This does not mean that a redesign of the meta-structures is a simple act based on theoretical insight. As a real change it has to be a process over time. There is a mutual contingency between individual morality, relationships, and structural setting. There is a mutual development of the self-identity of the sustainable person, the sustainability relations, and sustainable meta-structures over time—a kind of self-enforcing spiral of mutual improvement, embedded in the context of societal interaction and agreement. In this sense the redesign of meta-structures realizes itself by a process of change interrelated with the simultaneous development of individual morality and enforced by the communities within which it occurs.

10.3 Examples of Internal Structural Change: Biomimicry, Industrial Ecology, and Fair Trade

The change of the meta-structures according to the three guiding principles is not a mere abstract philosophical idea. We can find some recent trends in science, technology and the economy that have the potential for a fundamental change and reorientation. *Biomimicry*, for instance, is a new movement in science and engineering that aims to develop technological innovations by examining natural processes and characteristics and orienting technological solutions on them. It is the idea of learning from nature (Benyus 2002). A famous example is the development of artificial surfaces that mimic the surface structure and self-cleaning mechanism of the leave of the lotus. The basic principle mimicked here is a specific rough microstructure of small bumps, which prevents the adhesion of any dirt (Barthlott and Neinhuis 1997; Solga et al. 2007).

Another technological movement that entails some reevaluation of technology and the economy is *industrial ecology*. Industrial ecology studies overall material and energy flows related to goods, production processes, or industries. It particularly suggests the design of circular production and consumption processes with minimal energy and material in- and output. The archetypes for this design are ecosystems and their circular processes, and the ultimate aim is the compatibility and integration of economic-technological processes into the natural system (see Frosch and Gallopoulos 1989; Ayres and Ayres 2001).

An example from the sphere of the economy is *fair trade*. Fair trade suggests and implements an alternative model of global business. It entails an orientation of business on social and ethical values, and on relationships with partners, particularly in poor countries (see, e.g., Raynolds et al. 2007) Fair trade has become an increasingly successful business model, supporting both profit and advanced ethical standards in regard to global relationships.

These examples demonstrate that there are some ongoing bottom-up developments and changes in the fields of science, technology, and the economy. They provide some potential of new basic evaluations for the meta-structures. To some extent, the independence paradigm is replaced by a new paradigm of interdependence, and the paradigm of separation is replaced by one of integration and interrelation. However, there is still some lack of philosophical underpinning in these changes. For instance, biomimicry and industrial ecology are implicitly based on forms of naïve empiricism. It is argued that nature is a perfect structure, an ideal on which humans should orient themselves.[7] This assumes that ethical guidelines could be given empirically. Furthermore, this represents a kind of reversed pattern of domination: Instead of the domination of nature by humans inherent to modern science

[7] See, e.g., Isenmann (2003) for a more detailed discussion of the philosophical underpinning of industrial ecology.

and technology, humans now should subordinate themselves to nature. Such underpinnings are problematic. They do not take the topic of the is–ought distinction seriously enough and remain in the modern pattern of separation of human and nature, instead of developing a new understanding of this relationship. However, these approaches have the potential to contribute to a fruitful practical change of the meta-structures and their basic evaluations. They would gain much strength if combined with philosophical approaches such as sustainability ethics. For the successful redesign of meta-structures, it would be important to combine both bottom-up approaches working within the meta-structures and philosophical reflections.

10.4 Excursus: Rationality, Human Self-identity, and Meta-structures

For the current state of the meta-structures, the underlying modern concept of the human as a rational being is of particular relevance. To understand the meaning of this concept, we have to consider the meaning of the term *rational*. Something is rational if it is organized or generated according to *ratio*. The Latin term *ratio* has two main etymological sources: First, it is the translation of the Greek term *logos*; second, it has the meaning of *calculation*. The former denotes a rather broad range of abilities of the human mind for considerations and thought, the latter a formal mechanism of thinking. The modern meaning of rationality, however, refers more to the meaning of calculation: we call something rational if it is calculable or in a calculable order, and rationality is the ability to recognize the principles of calculation or the calculable order. This is the basis of the specific modern scientific and economic concepts of rationality. Scientific rationality refers to the calculability of the natural world. Economic rationality refers to the calculability of human behavior and the results of human interactions. By both, the human being feels able to calculate and control the world—i.e., natural and societal processes. In modern European thought, this idea of rationality is combined with the basic evaluation that it is good to strive for calculation and control of the world—that this will improve human well-being.

The modern idea of rationality has become increasingly dominant and is considered widely to be the sole principle of human thought and of the understanding of all phenomena. Consequently, it has also been extended to the human being itself, which now also becomes an object of rational calculation. The human being is considered to be an economic person whose behavior is calculable, as it is the result of a formal rational process. The human being also becomes an object of genetics, brain research, and neuroscience, which all suggest that the human being is completely determined by mechanisms that can be understood by scientific rationality and empirical recognition of causal laws. These recent developments in science have an increasing influence. The scientific consensus that the human being can be fully understood by scientific research as a rational, i.e., calculable mechanism, is growing in prevalence. With sufficient research, all ignorance and uncertainty, all seeming indetermination will supposedly be erased and the human being fully explained. This would be the ultimate fulfillment of the Baconian dream of complete recognition and control of the world. Even the most dubious and uncontrollable factor—the human being itself—would become a calculable and controllable factor. This would indeed allow the full determination and control of the world and the future.

This specific modern rational approach is, however, neither possible nor ethically desirable. The impossibility results from three aspects: First, it results from underlying epistemological assumptions that refer to the illusion of an infinite amount of ignorance and knowledge, erasable by rationality (see Sect. 7.1). Second, it results from the generalization of the specific modern interpretation of rationality. This ignores other forms of rationality and, with this, other principles of rational order, bringing-forth, action, etc. This becomes evident if we refer to the original meaning of *ratio* as a translation of the Greek term *logos*. According to Aristotle (2000: VI), there are several distinct abilities related to the logos, i.e., there are different forms of rationality and knowledge. Aristotle distinguished *episteme*, which is roughly similar to the rationality of calculation in the modern sense, from other types of rationality such as practical wisdom (*phronesis*), which guides human actions, and know-how (*techne*), which guides bringing-forth and production. Altogether, Aristotle distinguished five types of rationality, of which only *episteme* is an ability related to calculation. All other forms of rationality are virtues of the mind that are developed over time by personal experience and learning in the respective fields. This means that only a smaller part of the world is a calculable order and can be understood in this way. A larger part of phenomena, of human and social activity and structures, as well as of nature, is not calculable. Orientation and guidance in these parts require other abilities of the mind, other forms of rationality, such as practical wisdom, which is relevant for all social interactions, including the economy, or know-how, which is relevant for the whole field of production and bringing-forth, including technology. Third, the impossibility of a successful totality of the modern rationality approach results from its underlying tacit assumption about the identity of rational construction and being. This is the naïve assumption that there is no gap between the rational constructions by the human mind and the very existence of things. However, theory is not real being, and the ongoing and increasing surprises we face when applying science to reality might be more plausibly explained by an irreducible and essential gap between rational construction and reality than by concluding that we have not yet reached the perfect construction.

It is one question if the claim of totality of the modern rational approach is possible and another question if it is desirable, i.e., if we should choose such an approach. I hold that there are at least two crucial ethical arguments against it. First, this approach means a dangerous reduction of the human being and its rationality; second, it has a dangerous effect on the dynamics of the meta-structures. Both ultimately result in a dangerous misinterpretation of the meaning, and of the sources of orientation, of the human being—of human actions and relations. In regard to the first aspect, we have to recognize that there is an immanent philosophical problem in the modern rationality approach. The idea of complete rational determination of the world and the human being ultimately ignores the potential of human freedom and morality. The striving for an external, objective orientation by rational calculation and determination does not recognize the inner potential of the human being to provide orientation by its own will, reason (Vernunft), identity, or virtues. The rationality approach suggests that there is a mere external orientation of human thought and action, i.e., an orientation given by rational insights into the calculable structures of the world. This ignores and denies the potential of an inner orientation by the morality of the individual person. Second, in regard to the meta-structures, we are confronted with a contradiction between ideal and result. The ideal is a control of nature, society, and the human being by scientific and economic rationality. The result is an overall dynamic increase of the meta-structures and their complexity, which might be said to be completely out of control. Increasing side effects and surprises counteract the original idea of rational control. I already have argued that this is an ethical problem, because it impedes individual morality as well as the development of excellent sustainability relations (Sect. 10.1).

Modern science and technology particularly ignore this gap between ideal and result and try to do more research and technological applications to overcome increasing surprises, problems, and side effects. This means that they use the rationality approach, which is a cause of the gap, as a tool to reduce it. The control and determination paradigm is still applied, in spite of the fact that it is not working. Modern science and technology lack the tools to analyze their underlying concept of rationality and to redesign it. Within the meta-structures one can proceed only on their basis and assumptions. With this, the Baconian dream easily may become a nightmare. By increasingly trying to implement this ideal, science and technology increasingly fail; the paradigm of rational control leads to the out of control dynamics of the current meta-structures. The modern concept of rationality—and its specifications occurring in science and economics—is a problematic reduction, and we need a more encompassing concept of rationality in the context of the meta-structures. We need a concept of rationality that would allow for an inner orientation of the meta-structures. This means a concept of rationality that includes abilities of guidance of individual morality, such as, e.g., practical wisdom, or reason as defined by Kant. This is a prerequisite for a sensible concept and realization of overall orientation of the meta-structures in light of the development of the sustainable person and its relations. It is a prerequisite for properly understanding our place in the sustainability relations and for adequately redesigning the meta-structures.

References

Aristotle. (1995). *Politics* (Books I and II) (T. J. Saunders, Trans.). Oxford: Clarendon Press.

Aristotle. (2000). *Nicomachean ethics* (R. Crisp, Ed. & Trans.). New York: Cambridge University Press.

Ayres, R. U., & Ayres, L. W. (Eds.). (2001). *A handbook of industrial ecology*. Cheltenham: Edward Elgar.

Ayres, R. U., Bergh, J. C. J. M. van den, & Gowdy, J. M. (1998). *Viewpoint: Weak versus strong sustainability* (Tinbergen Institute Discussion Papers, 98-103/3). Amsterdam/Rotterdam: Tinbergen Institute.

Barthlott, W., & Neinhuis, C. (1997). Purity of the sacred lotus, or escape from contamination in biological surfaces. *Planta, 202*, 1–8.

Baumgärtner, S., Becker, C., Faber, M., & Manstetten, R. (2006). Relative and absolute scarcity of nature. Assessing the roles of ecology and economics for biodiversity conservation. *Ecological Economics, 59*, 487–498.

Becker, C. (2006). The human actor in ecological economics: Philosophical approach and research perspectives. *Ecological Economics, 60*, 17–23.

Becker, C. (2007). Die Mensch-Umwelt-Beziehung in den Wirtschaftswissenschaften. In T. Knopf (Ed.), *Umweltverhalten in Geschichte und Gegenwart* (pp. 212–227). Tübingen: Narr.

Becker, C. (2008). Thoreau's economic philosophy. *The European Journal of the History of Economic Thought, 15*, 211–246.

Becker, C. (2009). Logos und Wirtschaft bei Aristoteles. Ein dogmenhistorischer Beitrag zur Diskussion des ökonomischen Rationalitätsbegriffes. *Archives for Philosophy of Law and Social Philosophy, 95*, 523–539.

Becker, C. (2011). *Foundations of business ethics*. Considerations on the methodology of business ethics and the roles of philosophy and economics within (SSRN Discussion Paper). Available at: http://ssrn.com/abstract=1773174)

Becker, C., & Jöst, F. (2007). Krise und Reform des Sozialstaates. Eine wirtschaftsethische Analyse des ethischen Dilemmas der ökonomischen Sozialstaatsreformvorschläge. *Zeitschrift für Wirtschafts- und Unternehmensethik, 8*, 126–141.

Becker, C., Faber, M., Hertel, K., & Manstetten, R. (2005). Malthus vs Wordsworth: Perspectives on humankind, nature and economy. A contribution to the history and foundations of ecological economics. *Ecological Economics, 53*, 299–310.

Benyus, J. M. (2002). *Biomimicry: Innovation inspired by nature*. London: Harper Perennial.

Daly, H. (1977). *Steady-state economics*. San Francisco: W. H. Freeman.

Daly, H. (1996). *Beyond growth*. Boston: Beacon.

Freeman, R. E. (1984). *Strategic management: A stakeholder approach*. Boston: Pitman.

Freeman, R. E. (1994). The politics of stakeholder theory. *Business Ethics Quarterly, 4*, 409–421.

Frosch, R. A., & Gallopoulos, N. E. (1989). Strategies for manufacturing. *Scientific American, 261*(9), 94–102.

Georgescu-Roegen, N. (1971). *The entropy law and the economic process*. Cambridge/London: Harvard University Press.

Heidegger, M. (1977). *The question concerning technology and other essays*. New York: Harper & Row.

Homann, K. (2002). *Vorteile und Anreize. Zur Grundlegung einer Ethik der Zukunft*. Tübingen: Mohr.

Isenmann, R. (2003). Further efforts to clarify industrial ecology's hidden philosophy of nature. *Industrial Ecology, 6*(3/4), 27–48.

Meadows, D. H., Meadows, D. L., Randers, J., & Behrens, W., III. (1972). *The limits to growth*. New York: Universe Books.

Neumeyer, E. (2004). *Weak versus strong sustainability: Exploring the limits of two opposing paradigms* (2nd ed.). Cheltenham: Edward Elgar.

Ott, K., & Döring, R. (2008). *Theorie und Praxis starker Nachhaltigkeit* (2nd ed.). Marburg: Metropolis.

Petersen, T., & Faber, M. (2006). The concept of responsibility. In S. Baumgärtner, M. Faber, & J. Schiller (Eds.), *Joint production and responsibility in ecological economics* (pp. 225–237). Cheltenham: Edward Elgar.

Raynolds, L., Murray, D., & Wilkinson, J. (2007). *Fair trade: The challenges of transforming globalization*. London: Routledge.

Smith, A. [1776](2000). *The wealth of nations*. New York: The Modern Library.

Solga, A., Cerman, Z., Striffler, B. F., Spaeth, M., & Barthlott, W. (2007). The dream of staying clean: Lotus and biomimetic surfaces. *Bioinspiration & Biomimetics, 2*, 126–134.

Solomon, R. (1992). Corporate roles, personal virtues: An Aristotelean approach to business ethics. *Business Ethics Quarterly, 2*, 317–339.

van den Berg, J. C. J. M., & Hofkes, M. W. (1999). Economic models of sustainable development. In J. C. J. M. van den Berg (Ed.), *Handbook of environmental and resource economics* (pp. 1108–1122). Cheltenham: Edward Elgar.

Part IV
Toward an Encompassing Sustainability Research

So far, I have focused on the ethical dimension of sustainability and identified and analyzed the characteristics of this dimension in detail. The results of my ethical analysis have important implications for the overall possibilities of academics to address the issue of sustainability, and implications for the proper design of an adequate academic approach in particular. An appropriate academic approach to sustainability must be able to refer to all core aspects of sustainability—to the aspects of continuance, orientation, and relationships. The discussion so far has demonstrated that sustainability particularly entails an important and complex ethical dimension, which requires a new type of sustainability ethics.

However, although the issue of sustainability cannot adequately be addressed without ethical considerations, it should be clear that it cannot be addressed by an ethical approach *alone*. The issue of sustainability also has a crucial factual dimension that needs to be analyzed. This holds in particular for the aspect of continuance. We need knowledge about the conditions for maintaining different kind of systems, processes, or entities. Also, the sustainability relations are to some extent contingent on, and simultaneously influence, the factual conditions given by reality. To analyze the factual dimension of sustainability, science can make important contributions. Thus, an encompassing approach to sustainability must be an integrated approach that draws on the insights of both ethics and science.

Such an integrated approach, however, is a difficult task. I will argue that, in light of the ethical considerations surrounding science as a meta-structure (see Chaps. 7 and 10), this combination of science and ethics will require some re-understanding and redesign of science, and ultimately an integration of both in a new type of sustainability research. In the following, I will discuss some basics of the proper design of such an academic approach. My considerations on the design of a new type of sustainability research are a contribution to the ongoing debate about the role of science for addressing issues of sustainability, and to the increasing demand for a redefinition of the scientific input in discussions of sustainability.

Chapter 11
The Need for a New Type of Sustainability Research

The role of science and research in analyzing and solving sustainability issues has been discussed since the concept became prominent in the 1980s. There have been several calls for new ways of addressing sustainability by scientific research, for a new role or new engagement of science, and even for a new kind of sustainability science. These calls are based on different arguments for why one cannot simply proceed with traditional research in regard to sustainability issues.

Some recent contributions ask for a *sustainability science* (Kates et al. 2001; Clark 2003; Komiyama and Takeuchi 2006) and are supported by influential scientific groups and organizations such as the Initiative on Science and Technology for Sustainability (ISTS) and the International Council for Science (ICSU). The following main characteristics are suggested for a new sustainability science: it would integrate several scientific disciplines, relate to both local and global levels, organize research as a process of cooperation between science and societal groups, and transfer knowledge to (local) societies; it would be problem- and action-driven, and it would be oriented toward the issues and problems of particular societies (Kates et al. 2001; ICSU 2002; Clark 2003; Komiyama and Takeuchi 2006).

The reasons given for such a model of sustainability science are generally derived from some characteristics (often tacitly) ascribed to sustainability issues: It has been recognized that in many cases environmental and sustainability issues have originally been defined by societal groups and in regard to practical contexts of action. With this, the issues neither fit into a specific disciplinary scientific framework, nor to the general orientation of science on cognitive interests rather than on action interests (Clark et al. 2004; Baumgärtner et al. 2008). It has been argued that the complex interplay between societal decisions and actions on the one hand and environmental changes on the other, which is typical for environmental and sustainability issues, requires a cooperation of several disciplines to analyze integrated social-environmental systems and their dynamics as a whole (Kates et al. 2001; ICSU 2002; Clark 2003). In addition, this interplay is said to require a new relation of science and society, one that would more adequately and robustly take into account the origin of sustainability and environmental issues in societal definitions, actions, and evaluations (Clark 2003). Furthermore, it has been recognized that

C.U. Becker, *Sustainability Ethics and Sustainability Research*,
DOI 10.1007/978-94-007-2285-9_11, © Springer Science+Business Media B.V. 2012

sustainability and environmental issues have long-term and global aspects, and that there are complex interrelations between global actions and local effects, as well as between contemporary actions and future results (Kates et al. 2001). Sustainability science therefore must have the ability to analyze the long-term dynamics of integrated systems and the interdependence of local and global levels of action.

The recent calls for a new sustainability science have several forerunners and parallel other similar concepts for redefining scientific research, such as *post-normal science* (Funtowicz and Ravetz 1993, 1994, 2003), *mode2 science* (Gibbons et al. 1994), *transdisciplinary research* (Thompson Klein et al. 2001; Hirsch Hadorn et al. 2008), the call for a "*new social contract for science*" (Lubchenco 1998; ICSU 2002), and the movement of *ecological economics* (Daly 1980, 1996; Costanza 1991; Baumgärtner et al. 2008).

Overall, one can identify four main features of sustainability research on which all these approaches and the discussions so far have focused: (1) the integration of several disciplines, (2) the integration of science and society, (3) the reference to localness, time, and uncertainty, and (4) action and problem orientation. All these aspects are without doubt of importance for the design of sustainability research. However, I hold that the discussions so far have concentrated too much on the adequate study of the factual aspects of sustainability—on the integrated study of systems and their dynamics and on the determination of factual parameters and conditions of continuance—and on epistemological issues and aspects of knowledge integration. With this, the ethical dimension of sustainability has been neglected, and important implications for sustainability research have not been recognized. In particular, the ethical meaning of science itself in regard to sustainability has rarely been considered so far and has not been adequately taken into account in the various attempts to develop requirements for a new sustainability research. Few approaches have considered the relevance of normative aspects (e.g., Daly 1980; Kates et al. 2005), but neither have fully identified and analyzed the ethical dimension of sustainability, nor drawn conclusions for the role of academics in addressing sustainability issues. Often, the normative dimension of sustainability has been reduced to empirical issues concerning the existing values and interests of societal groups (e.g., Kates et al. 2005: 16f). The overall disregard of the full meaning of the ethical dimension of sustainability has led to serious shortcomings in the discussion and design of sustainability research.

In the following, I will present considerations for the design of a new sustainability research, which are explicitly based on the analysis of the ethical dimension of sustainability provided so far in this book. I hold that a discussion of an adequate academic approach can only be grounded in the characteristics of the subject matter—the issue of sustainability—itself. As I have demonstrated, one crucial characteristic of the subject matter is its specific ethical dimension. Thus, in the following, I refer to the consequences of the specific characteristics of the ethical dimension for the design of an adequate academic approach to sustainability.

I have argued so far that the issue of sustainability has an inherent ethical dimension, and I have analyzed this dimension in detail. I have particularly demonstrated that science itself, as a meta-structure, is part of the ethical dimension of sustainability.

Science is based on certain fundamental assumptions and evaluations and affects the sustainability relations in several ways. In this respect, I gave an ethical critique of the meta-structure science, as well as general ethical guidelines for the redesign of meta-structures (Chap. 10). Thus, on the one hand, from an ethical perspective, science itself is part of the issue of sustainability. On the other hand, science can be an important tool for the analysis of the issue of sustainability. Science provides methods for analyzing the continuance of specific systems, entities, and processes, and with this, it provides important approaches to the continuance aspect of sustainability. Science constitutes an important approach to the factual aspects of sustainability, particularly of the sustainability relations. For instance, the human–nature relationship is not just actualized and governed by ethical categories or by the way humans relate themselves deliberately to nature, but also by biological facts and necessities that can be analyzed by science.[1]

Because the sustainability issue has ethical and factual dimensions, both of which are crucial, sustainability research must integrate both ethics and science in an encompassing academic approach to sustainability. This is the first requirement for a sustainability research: *it must encompass both ethics and science*. That this requirement cannot be fulfilled by simply adding the two should be clear from the analysis in Part III. We cannot just apply science as it is, with its inherent normative and evaluative assumptions, its paradigms of human independence, control, and growth, and its nonreflected dynamics of increasing complexity. To approach sustainability, we need to reflect on the double role of science as a *tool* for analyzing sustainability and as a *part* of the sustainability issue. We need to reflect on science as being a meta-structure relevant for sustainability and affecting the sustainability relations; we need to reflect on the inherent normativity of science; we need to design and orient science in light of the sustainability relations; we need to refer to the production and transfer of knowledge as a whole as an issue of continuance and sustainability. Ethics must, therefore, be an integral part of sustainability research that allows for internal self-reflection on the normative assumptions of science, as well as the ethical meaning of the overall constitution and dynamics of science in regard to sustainability. This holds in general terms for science as a whole, as well as for particular disciplines such as physics, biology, or economics, each of which entail a set of more specific basic assumptions and evaluations that need to be explicitly reflected on and ethically discussed in light of sustainability issues.

The second requirement for sustainability research therefore is a *coherent integration of science and sustainability ethics,* which ensures critical self-reflection of scientific approaches and underlying assumptions, as well as the adequate overall

[1] This does not mean that science is the only way to refer to the factual dimension of sustainability. Many cultures have based their reference to this dimension on other types of recognition and knowledge, for instance, on personal experience and know-how, and on transfer of this knowledge from person to person. However, considering the current status quo, the crucial role of science for the understanding and constitution has to be taken into account. It seems to be more fruitful to include and redesign this approach than to completely deny it (see also Chap. 13).

orientation of sustainability research. Such an integrative approach particularly needs to avoid contradictions between scientific approaches and sustainability ethics. For this, it should be based on an adequate redesign of science in accordance with the general guidelines for restructuring the meta-structures. That is a redesign in regard to stability and simplicity of the meta-structures, new basic evaluations underlying the meta-structures, and an encompassing underlying concept of the human being as a fundamentally relational being (see Sect. 10.2).

These guidelines have three implications for developing sustainability research. First, we have to ensure an inner stability and simplicity of scientific knowledge production and transfer and to avoid uncontrolled increase of complexity and fragmentation of science. I will argue that this can partly be supported by organizing sustainability research as a specific type of interdisciplinary research (Sect. 12.1). Secondly, we need a new underlying basic evaluation for science. The paradigm of independence and control has to be replaced by a new paradigm of interdependence and interrelation. The growth paradigm, i.e., the simple assumption that more knowledge is good, has to be replaced by an overall orientation of science toward the constitution of the sustainability relations. I will argue that this can partly be implemented by designing sustainability research as a certain type of transdisciplinary research (Sect. 12.2). The third element, which is a crucial complement to these more organizational aspects of a new sustainability research, is the personal identity and competencies of the researcher. The researcher within sustainability research needs to possess a set of specific cognitive and communicative skills to actualize this new type of inter- and transdisciplinary research, including the ability for ethical and methodological reflection on her own scientific discipline and approaches. Moreover, I will argue that the researcher needs to develop a certain personal identity and individual morality: she has to be able to integrate her identity as a scientific person into her overall identity as a sustainable person. This personal identity is a crucial factor for the ethical orientation and new basic evaluation of sustainability research (Chap. 13). All three aspects together provide the basis for a new type of sustainability research that allows for an adequate integration of scientific research and sustainability ethics.

An encompassing academic approach to sustainability will clearly distinguish itself from the established disciplinary research. It will be a new type of research that integrates ethical and factual analyses, and entails the potential for self-reflection on its own role and ethical self-orientation in regard to sustainability. It is based on a restructuring and reorientation of science—which means a reorientation of its basic assumptions and evaluations, and its inner dynamic—in reference to philosophical tools and sustainability ethics. In the following, I will discuss the main features of this new type of sustainability research in detail, i.e., its characteristic as a kind of inter- and transdisciplinary research, the requirements concerning the capabilities and personal identity of the researcher, and the role of philosophy within it.

References

Baumgärtner, S., Becker, C., Frank, K., Müller, B., & Quaas, M. (2008). Relating the philosophy and practice of ecological economics. The role of concepts, models, and case studies in inter- and transdisciplinary sustainability research. *Ecological Economics, 67*, 384–393.

Clark, W. C. (2003). *Sustainability science: Challenges for the new millennium*. An address on 4 September at the official opening of the Zuckerman Institute for Connective Environmental Research and the Third Sustainability Days, September 4–10, 2003, University of East Anglia, Norwich. http://sustainabilityscience.org/ists/docs/clark_zicer_opening030904.pdf. Accessed April 15, 2011.

Clark, W. C., Crutzen, P. J., & Schellnhuber, H. J. (2004). Science for global sustainability. Toward a new paradigm. In H. J. Schellnhuber, P. J. Crutzen, W. C. Clark, M. Claussen, & H. Held (Eds.), *Earth system analysis for sustainability* (pp. 1–28). Cambridge: MIT Press.

Costanza, R. (1991). *Ecological economics: The science and management of sustainability*. New York: Columbia University Press.

Daly, H. (Ed.). (1980). *Economics, ecology, ethics. Essays toward a steady-state economy*. San Francisco: W. H. Freeman.

Daly, H. (1996). *Beyond growth*. Boston: Beacon.

Funtowicz, S., & Ravetz, J. (1993). Science for the post-normal age. *Futures, 25*, 739–755.

Funtowicz, S., & Ravetz, J. (1994). The worth of a songbird: Ecological economics as a postnormal science. *Ecological Economics, 10*, 197–207.

Funtowicz, S., & Ravetz, J. (2003). Post-normal science. In International Society for Ecological Economics (Ed.), *Online encyclopedia of ecological economics*. http://www.ecoeco.org/pdf/pstnormsc.pdf. Accessed April 15, 2011.

Gibbons, M., Limoges, C., Nowotny, H., Schwartzman, S., Scott, P., & Trow, M. (Eds.). (1994). *The new production of knowledge*. London: Sage.

Hirsch Hadorn, G., et al. (Eds.). (2008). *Handbook of transdisciplinary research*. New York: Springer.

International Council for Science (ICSU). (2002). *Science and technology for sustainable development* (Series on Science for Sustainable Development, No. 9). Paris: ICSU.

Kates, R. W., et al. (2001). Sustainability science. *Science, 292*, 641–642.

Kates, R. W., Parris, T. M., & Leiserowitz, A. A. (2005). What is sustainable development? Goals, indicators, values, and practice. *Environment: Science and Policy for Sustainable Development, 47*(3), 8–21.

Komiyama, H., & Takeuchi, K. (2006). Sustainability science: Building a new discipline. *Sustainability Science, 1*, 1–6.

Lubchenco, J. (1998). Entering the century of environment: A new social contract for science. *Science, 279*, 491–497.

Thompson Klein, J., Grossenbacher-Mansuy, W., Häberli, R., Bill, A., Scholz, R. W., & Welti, M. (Eds.). (2001). *Transdisciplinarity: Joint problem solving among science, technology, and society. An effective way for managing complexity*. Basel: Birkhäuser.

Chapter 12
Inter- and Transdisciplinarity

Sustainability research requires a new scientific self-identity and a new organizational and institutional framework of research. I will refer to the existing concepts of interdisciplinarity and transdisciplinarity to discuss some main principles for the design of sustainability research. With this, I both refer and contribute to recent discussions about new forms of science and research, particularly new forms of environmental research (see Becker and Baumgärtner 2005). However, I will provide a specific interpretation and definition of the concepts of inter- and transdisciplinarity, which currently have a rather broad and sometimes vague meaning, to make them fruitful for the design of the new type of sustainability research.

12.1 Interdisciplinary Integration of Sciences and Sustainability Ethics

It has already been recognized that some important issues, such as environmental and sustainability issues, require an interdisciplinary analysis. The call for interdisciplinarity results from the understanding that these issues are complex ones and cannot be covered fully by a single scientific discipline. Modern science is fragmented into several disciplines and subdisciplines, each of which can deal only with a small part of complex issues. For instance, global warming requires the analysis of the atmosphere, oceans, economy, social impacts, and so on. Each of these aspects refers to a different scientific discipline, such as physics, chemistry, biology, economics, etc. The obvious implication is that we need to "put together,"—to combine—all relevant scientific disciplines and approaches in order to analyze and understand the issue fully. This idea of combining several scientific disciplines is the basic meaning of the concept of *interdisciplinarity*.

However, there is a range of possibilities to realize and organize a combination of different scientific disciplines, i.e., an interdisciplinary cooperation. A rather simple way of interdisciplinary cooperation is to simply add the results of different

C.U. Becker, *Sustainability Ethics and Sustainability Research*,
DOI 10.1007/978-94-007-2285-9_12, © Springer Science+Business Media B.V. 2012

disciplinary analyses of a specific issue, for instance, in a volume or expertise. Another type of interdisciplinary cooperation would be a division of labor between the disciplines, in which one discipline uses the results of the other as an input into its own analysis. A further version of interdisciplinarity would be to use a certain method of another discipline within one's own research.[1]

All these types of interdisciplinary cooperation between scientific disciplines are somehow insufficient and can run into several problems (Becker and Baumgärtner 2005; Baumgärtner et al. 2008). The main reason is that each discipline is based on a set of basic concepts and assumptions, and defines and approaches its subject matter in a specific way on this basis. The basic concepts and assumptions of different disciplines do not always fit together without friction. Therefore, if results or methods from different disciplines are borrowed by others or simply put together, hidden and unrecognized frictions or even contradictions can result. For instance, the assumptions about the human being underlying economics are rather specific and will not always harmonize with the explicit or implicit understanding of the human being in other sciences (see Becker 2006). Another example is the implicit understanding of nature in economics, which does not fit with the understanding of nature in biology or ecology (Becker 2007; Baumgärtner et al. 2006). Therefore, a successful and well-founded interdisciplinary cooperation would require integration, or at least an explicit discussion, at the level of basic concepts and assumptions. It would require a common, coherent, and consistent definition of both the subject matter and of the adequate approach to it. This, however, amounts to a more demanding type of interdisciplinary research that starts with the development of a framework of integrated concepts and methods.

Such an *integrated type* of interdisciplinary research would provide more than an encompassing and coherent approach to sustainability issues. It is not just about knowledge integration. It would provide a framework for a fundamental integration of the very *process* of scientific knowledge production. It would relate the different disciplines and focus them on a common basis and set of concepts and methods. This integrated type of interdisciplinary research could in particular be a means for counteracting the inter-scientific driving forces of specialization and expansion of knowledge, the ongoing process of increasing fragmentation and complexity within traditional disciplinary research. With this, it can be a means for stabilizing and balancing the dynamics of modern science as a whole, which particularly would stabilize knowledge transfer to future generations.

By this function, the specific type of integrated interdisciplinary research is one important organizational feature of the new type of sustainability research, because it particularly is a means for redesigning the dynamics of modern science and for promoting the integrity, stability, and continuance of science as a whole, which is a

[1] For a more detailed discussion of different types of interdisciplinary research, see Baumgärtner and Becker (2005) and Baumgärtner et al. (2008).

crucial aspect of the sustainability issue and its adequate analysis (see Chaps. 7 and 10). It should be clear, however, that such a type of interdisciplinary research is not only an intellectual challenge, but also a challenge for the disciplinary structure and organization of modern science. It would require overcoming the disciplinary organization of science, and particularly the disciplinary evaluation processes, which often do not allow for, or at least underappreciate, interdisciplinary research.

However, as I suggested earlier, the quality of interdisciplinarity becomes most relevant and fruitful for sustainability research when its conception is expanded to a more fundamental kind of integration: the integration of ethical and factual analysis. As argued above, this integration cannot be done by a simple putting-together of ethics and science, or of the results of both. Science itself and its design and procedures are ethically relevant to the issue of sustainability. Science as a whole and each specific discipline in particular entails various ethical relevant assumptions and evaluations, which are relevant in regard to sustainability issues. Thus, ethical considerations need to be integrated into scientific research to enable science to reflect on its own assumptions and evaluations, and to orient its research and own development properly toward sustainability issues. Only such a full integration of ethics and scientific research will lead to a complete recognition and adequate consideration of the ethical dimension of sustainability, and of the ethical requirements for scientific research itself. The integration of science and sustainability ethics would result in sustainability research that can approach the factual and ethical aspects of sustainability in a consistent and encompassing manner.

Together, the integration of ethical and factual analysis *and* the interdisciplinary integration of different scientific disciplines ultimately fully define the concept of interdisciplinary integration I consider to be a basis for sustainability research. This kind of interdisciplinary integration is a demanding project. It requires the overcoming of traditional organization of disciplinary research, and it is in contrast to the traditional restriction of modern science to mere factual analysis. Such a new understanding of research requires some further reflections on the self-identity and organization of sustainability research. It has also to be grounded on further crucial characteristics of sustainability research, namely on transdisciplinarity and the personal identity of the researcher, which I will discuss in the following chapters.

Overall, the specific interdisciplinarity of sustainability research is a challenge. It defines a new type of research with a demanding integration of scientific disciplines and sustainability ethics. It requires the capacity for self-reflection on the fundamentals and structure of modern science as a whole, as well as on the basics of specific disciplines. This ultimately is more than a mere ethical issue. It is a philosophical issue that requires a reference to the philosophy and history of science, as well as the philosophy and history of particular sciences. Philosophical reflection must provide the basis for the integration of disciplinary research and sustainability ethics and, with this, provide the basis of the interdisciplinary organization of a new type of sustainability research.

12.2 Transdisciplinary Integration of Research into the Sustainability Relations

A crucial requirement for sustainability research is the constitution of a new basic evaluation for it, and its proper orientation in the context of the sustainability relations. In the following discussion of the respective design of its self-identity and organizational framework, I refer to the term *transdisciplinarity*, which has already been established in methodological discourses to denote a certain alternative type of research.

The term *transdisciplinarity* commonly refers to the vision of a type of scientific research that relates to society in a new way (Thompson Klein et al. 2001; Hirsch Hadorn 2002; Hirsch Hadorn et al. 2006, 2008). In a transdisciplinary research, the formulation of research questions, the research process, and the application of results are all done by science and society together—by the strong interaction of science and relevant social groups.[2] This conception of a new type of research aims to overcome the current design of science, which is often driven by internal incentives and research questions, and is not explicitly oriented on crucial societal issues, such as environmental or sustainability issues. It is the idea of referring science to society and its issues in a new and explicit way, and making societal issues the main orientation of scientific research.

The discussion of transdisciplinarity addresses the fundamental issue of the role of science within society. It actually is more of a *trans-scientific* issue than a transdisciplinary issue. Within the terminology of this book, I reinterpret the striving for transdisciplinarity as the striving for a redesign of the meta-structure of science in light of its role within society. Transdisciplinarity represents the demand that science replace its old basic evaluations and orientation, both of which are said to have caused an alienation of science and society, with an explicit orientation toward problems and values as defined by societal groups. Transdisciplinarity represents the demand that we replace the basic evaluation that science will automatically improve society with a basic evaluation that is based on societal evaluations and decisions. That means that decisions about what research is done, in what way, and for what purpose, are no longer merely inner-scientific decisions, but are decisions made together by science and society. The implicit "old social contract" underlying modern science—that more scientific research and knowledge will automatically lead to an improvement of society—is replaced by a "new social contract" of explicit relationship between science and society. This in particular means that scientific research is now based on societal evaluations about what problems are most crucial, what research is most important, etc., and these evaluations become part of the new basic orientation of a transdisciplinary science.

In addition, transdisciplinarity counteracts the dominance of scientific knowledge and the claim of totality of modern science in regard to knowledge and knowledge

---

[2] The concept of transdisciplinarity is also used in a variety of other meanings. Sometimes it is not defined as an interconnection of science and society, but rather as a certain type of interdisciplinarity, e.g., by Costanza et al. (1998) and Mittelstraß (2001).

production. Transdisciplinary research explicitly refers to existing knowledge within society, which includes knowledge based on personal experience, know-how, culturally transmitted knowledge, etc., and aims to incorporate those types of knowledge into the definition and analysis of its subject matter. This means an extended type of knowledge integration, which has its specific difficulties and challenges (see also Hirsch Hadorn et al. 2008).

The discussion of transdisciplinary research has also led to suggestions for the realization and actualization of such a type of science. This particularly includes means for participation of individuals and societal groups in research processes (Elzinga 2008) and ideas for designing adequate processes and institutions for communication between scientists and societal groups (see Elzinga 2008). In this respect one could also say that the idea of transdisciplinarity entails a conception of democratization of scientific research.

The idea of transdisciplinarity is fruitful for developing a new type of sustainability research. The idea provides some suggestions for a new embedment of science into society and, with this, gives stimulus for establishing a new basic evaluation and orientation of scientific research. However, it is not just the role of science within society that is of relevance for a new type of sustainability research, but it is the *overall role of science and research within the sustainability relations* as a whole. Therefore, for the purpose of designing a new type of sustainability research, I suggest to *extend* and redefine the concept of transdisciplinarity. As a characteristic of sustainability research, I define transdisciplinarity as redesign and reorientation of the meta-structure science with regard to the context of *all* three sustainability relations. This is a broader understanding of transdisciplinarity, which means not just a new *social* contract for science to reorient science on society, but rather a *new sustainability contract for science* to reorient science toward the three sustainability relations—toward society, future generations, and nature.

To understand and design sustainability research in this way as transdisciplinary research enables sustainability research to reflect on the embedment of science in the sustainability relations and to actualize and orient itself within them. Transdisciplinarity means a fundamental integration of research into the development of the sustainability relations. However, the realization of transdisciplinary sustainability research is a challenge that goes beyond the difficulties discussed so far in regard to transdisciplinarity. It can be supported, but not fully realized and established, by participation and communication strategies. The ideas of participation and communication of the researcher with societal groups developed in the established conception of transdisciplinarity are fruitful for relating sustainability research to contemporaries. In this case, the ethical orientation of sustainability research results from its reflected actualization as a means for the constitution and development of societal relationships, particularly as a means for the analysis and solution of societal problems.

However, sustainability research has to reflect on the role of science and research as means of constituting, maintaining, and governing all three sustainability relations—the relation with contemporaries, future generations, and nature. For this, concepts such as participation or communication are of limited relevance, as of course neither future generations nor nature are able to participate in research processes. We rather need an ethically informed sustainability research, which is guided by sustainability ethics. Definition of issues and research questions, choice of approaches, and interpretation of results must be done in the light of their relevance and meaning for all three sustainability relations together.

Such a transdisciplinary sustainability research is ethically informed and oriented toward the development of a sustainable future. The ethical orientation of this research is established by the integration of sustainability ethics into the research processes. However, the idea here is not that specialists of sustainability ethics are responsible for the ethical orientation of sustainability research. Rather, this ethically informed research is essentially based on individual morality, ethical competence, and specific capabilities of each researcher and scientist involved in sustainability research. The integrative challenge of inter- and transdisciplinarity is not merely an organizational issue, but requires specific competencies and capabilities of each researcher involved.

References

Baumgärtner, S., & Becker, C. (Eds.). (2005). *Wissenschaftsphilosophie interdisziplinärer Umweltforschung*. Marburg: Metropolis.

Baumgärtner, S., Becker, C., Faber, M., & Manstetten, R. (2006). Relative and absolute scarcity of nature. Assessing the roles of ecology and economics for biodiversity conservation. *Ecological Economics, 59*, 487–498.

Baumgärtner, S., Becker, C., Frank, K., Müller, B., & Quaas, M. (2008). Relating the philosophy and practice of ecological economics. The role of concepts, models, and case studies in inter- and transdisciplinary sustainability research. *Ecological Economics, 67*, 384–393.

Becker, C. (2006). The human actor in ecological economics: Philosophical approach and research perspectives. *Ecological Economics, 60*, 17–23.

Becker, C. (2007). Die Mensch-Umwelt-Beziehung in den Wirtschaftswissenschaften. In T. Knopf (Ed.), *Umweltverhalten in Geschichte und Gegenwart* (pp. 212–227). Tübingen: Narr.

Becker, C., & Baumgärtner, S. (2005). Einleitung. In S. Baumgärtner & C. Becker (Eds.), *Wissenschaftsphilosophie interdisziplinärer Umweltforschung* (pp. 9–23). Marburg: Metropolis.

Costanza, R., Cumberland, J., Daly, H., Goodland, R., & Norgaard, R. (1998). *An introduction to ecological economics*. Boca Raton: St. Lucie Press.

Elzinga, A. (2008). Participation. In G. Hirsch Hadorn et al. (Eds.), *Handbook of transdisciplinary research* (pp. 345–359). New York: Springer.

Hirsch Hadorn, G. (2002). Unity of knowledge in transdisciplinary research for sustainability. In UNESCO-EOLSS Joint Committee (Ed.), *EOLSS—Encyclopedia of life support systems*. Oxford: EOLSS Publishers.

Hirsch Hadorn, G., Bradley, D., Pohl, C., Rist, S., & Wiesmann, U. (2006). Implications of transdisciplinarity for sustainability research. *Ecological Economics, 60*, 119–128.

Hirsch Hadorn, G., et al. (Eds.). (2008). *Handbook of transdisciplinary research*. New York: Springer.

Mittelstraß, J. (2001). *Wissen und Grenzen. Philosophische Studien*. Frankfurt: Suhrkamp.

Thompson Klein, J., Grossenbacher-Mansuy, W., Häberli, R., Bill, A., Scholz, R. W., & Welti, M. (Eds.). (2001). *Transdisciplinarity: Joint problem solving among science, technology, and society. An effective way for managing complexity*. Basel: Birkhäuser.

Chapter 13
Capabilities and Personal Identity of the Researcher

The new type of sustainability research makes high demands on the individual researcher, her capabilities, and personal identity. The individual researcher is a key element for the development and orientation of sustainability research. The characteristics of the issue of sustainability and the specific inter- and transdisciplinary design of sustainability research result in specific requirements for the sustainability researcher in regard to her (1) cognitive capabilities, (2) communication skills, and (3) personal morality and ethical competencies.[1] This goes far beyond the mere requirements of analytical and scientific excellence usually made on a traditional scientist, which are, of course, also requirements for the sustainability researcher.

1. *Cognitive capabilities.* Sustainability research is a certain type of inter- and transdisciplinary research. The interdisciplinary integration requires the cognitive capability to critically reflect on the basic assumptions, methods, and concepts of one's own as well as of other disciplines. It requires the capability not just to apply methods within a given theory and set of concepts, but to do conceptual work, theory building, definition of integrative basic concepts, and development of integrated methods. This requires some knowledge of the differences and specifics of scientific disciplines in general, including their historical development, self-identity, subject matter, scientific standards, etc. Moreover, it requires knowledge of the theories, main methods, and basic concepts of the specific disciplines with which one is cooperating, as well as the ability to critically analyze them. Finally, because ethics is a crucial component of interdisciplinary sustainability research, the researcher within must be able to recognize and understand ethical issues, particularly ethical issues of sustainability and scientific research, including inherent normative and evaluative elements of science and specific scientific disciplines. All these requirements do not mean that everyone has to know everything, but that everyone involved in sustainability research should be able to

[1] See also Faber (2008) and Baumgärtner et al. (2008).

C.U. Becker, *Sustainability Ethics and Sustainability Research*,

DOI 10.1007/978-94-007-2285-9_13,

understand all relevant aspects of the issue and the basics of the other disciplines involved. This is a basic requirement for successful integrated interdisciplinary research.

The transdisciplinarity of sustainability research requires the researcher to develop the capability to deal with societal concepts and descriptions of problems and transfer them into scientific research, and vice versa. She needs the capability to analyze the meanings of societal concepts carefully and to make sure that they will be adequately transferred into scientific concepts, and vice versa. Unconsidered simple replacement of societal concepts by existing scientific concepts should be avoided, as should larger frictions between societal and scientific formulations of a certain problem. For instance, there are often major differences in the definitions, descriptions, and normative prescriptions of an environmental problem by societal groups, environmental economics, and ecology, and these differences should be carefully considered. In a similar vein, different sciences and societal groups may use the same concept, but with very different meanings. Concepts such as sustainability are understood and defined in rather different ways in society, politics, natural sciences, and economics. Careful conceptual work is therefore necessary for the adequate consideration of societal concepts and problem descriptions in sustainability research, to facilitate meaningful dialogue among the sciences, ethics, politicians, and society, and to develop a transdisciplinary research process. The sustainability researcher needs the capability to adequately do this work.

Furthermore, the researcher needs the capability to refer to other forms of knowledge, such as knowledge based on personal experience, concrete skills, know-how, culturally transmitted knowledge, etc., and the researcher should be able to refer to forms of personal experience such as emotional or intuitive experiences, which are relevant for a full understanding and development of the sustainability relations, particularly the human–nature relationship. The sustainability researcher also needs the capability to integrate the scientific approach and knowledge as one specific type of approach and knowledge with a broader conception of human recognition and knowledge.

2. *Communication skills.* The challenges of inter- and transdisciplinary research also require specific communication skills of the sustainability researcher. This type of research is not merely a cognitive challenge that can be met by the single researcher, but a challenge to be met by the interaction of many researchers from different disciplines and laypersons from societal groups. This interaction requires a specific yet diverse set of communication skills. The researcher must possess the ability to explain the assumptions, concepts, methods, and problem descriptions of her own discipline to researchers from other disciplines as well as to laypersons, and she must also be capable of receiving and grasping the assumptions, concepts, methods, and problem descriptions of other disciplines and of laypersons.

   This is more a difficult task than it may seem at first sight. Concepts that are self-evidently used in a scientific discipline are sometimes rather difficult to explain because they are not always explicitly defined, but are common knowledge within the discipline, e.g., the concept of utility in economics. In other words, within a given discipline, a word's common usage, context, and conceptual associations can be greater aspects of its meaning than a straightforward definition. Another problem is the high level of formal language and analytical analysis in many sciences. There may be a clear definition of concepts and a proper argument, but this often occurs in a formal and analytical way that makes it difficult to translate it in everyday language. This holds for most natural sciences and also for economics. To some extent the communication between disciplines and with society can be supported by organizational instruments, e.g., by adequate design and moderation of inter- and transdisciplinary meetings and discussions (see also Mieg 2003). However, the capability of the researcher for communication within this context remains a necessary precondition for successful inter- and transdisciplinary research.

The cognitive capabilities and communication skills that are required for sustainability research are demanding, but a necessary precondition for its success. Without them, successful inter- and transdisciplinary research is not possible. We could, for instance, bring together the most distinguished experts of all disciplines relevant for analyzing a certain problem, but without the capabilities described above, they would be unable to understand each other and bring together their disciplinary perspectives and knowledge in a fruitful way. There would be no integration, no common result, and no solution to the problem.

3. *Personal morality and ethical competencies.* Sustainability has an ethical dimension and sustainability research has to meet this dimension. The researcher within must be able to refer to and adequately deal with the ethical dimension of sustainability and to reflect on the role of science and its own role in regard to the sustainability relations. One of the goals of sustainability research is the proper development of science as a structure that supports the constitution and governance of the sustainability relations in an excellent way. The crucial reference point for the ethical dimension and the structural redesign of science is the sustainable person. The ethical orientation of science and sustainability research is, therefore, not merely an organizational and structural issue, but depends on the individual self-identity and ethical competencies of the researchers involved. The basic evaluations of sustainability research cannot be designed and defined in an exclusively theoretical and abstract way. They have to be developed and actualized by the participating researchers. The researcher within sustainability research has the challenge of developing an identity as sustainable person and of integrating her identity as a researcher within it. This means that she should understand her scientific identity as part of her identity as a sustainable person. She has to develop and embed her scientific identity respectively.

The implementation of a new sustainability research is a process of mutual development of inter- and transdisciplinary organizational structure and personal identity of the researcher. The transdisciplinary design is a formal conception that would allow the researcher to develop her identity as a sustainable person and integrate her scientific identity within. However, this requires that the researcher commits to developing, actualizing, and supporting this formal frame through the development of her personal identity and morality. Only this mutual supportive development of individual morality and organizational framework will result in a proper ethical orientation for sustainability research. Inter- and transdisciplinary organization helps to structure sustainability research and embed it into the sustainability relations, as the respective personal capabilities will help to relate the researchers to each other and to the sustainability relations.

Also, the sustainability researcher needs to develop and cultivate the willingness and openness to critically reflect on her own discipline and methods. This includes the reflection on the (normative) assumptions underlying her own research, and on the possibilities and limits of her own discipline to refer to certain societal issues. This not only requires analytical and scientific skills and knowledge, but also skills of ethical and methodological reflection. It is also a matter of personal attitude and identity of the sustainability researcher, who needs to develop a certain critical distance to her own discipline, and a positive attitude toward critical discussion of the limits of her own discipline and its methods. Distance to her own doing and role, openness to critical reflection and willingness for critical discussion can also be considered as specific "research virtues" that the sustainability researcher needs to cultivate in addition to her virtues as being a sustainable person.

References

Baumgärtner, S., Becker, C., Frank, K., Müller, B., & Quaas, M. (2008). Relating the philosophy and practice of ecological economics. The role of concepts, models, and case studies in inter- and transdisciplinary sustainability research. *Ecological Economics, 67*, 384–393.

Faber, M. (2008). How to be an ecological economist. *Ecological Economics, 66*, 1–7.

Mieg, A. (2003). Interdisziplinarität braucht Organisation! *Umweltpsychologie, 7*(2), 32–52.

Chapter 14
The Role of Philosophy for Sustainability Research

Philosophy plays a crucial role for sustainability research in several ways. First, the definition and analysis of the ethical dimension of sustainability is a subject of ethics, and with this, belongs to the realm of philosophy. However, the ethical dimension of sustainability cannot be analyzed completely within the field of ethics, as it requires the reflection of the ethical role of science as well. Sustainability research must consider science as both a tool for approaching the issue of sustainability and part of the issue. Sustainability research thus needs to develop the ability and methods to reflect on its double role and to analyze and orient its own role within the sustainability relations. This is a further philosophical task, which can particularly be supported by epistemology and philosophy of science. With this, philosophy enables researchers for methodological reflections, and for discussion of the foundations of science, its inherent normative assumptions, and its ethical meaning in light of sustainability.

In addition, philosophy provides fundamentals for the inter- and transdisciplinary design of sustainability research. The interdisciplinary integration of sciences and ethics requires more specific philosophical analyses of the self-identity, concepts, methods, basic assumption, evaluations, and implicit normative aspects of particular sciences, the discussion of their relationships, and their integration. This cannot be done by the methods of the respective sciences alone, but requires a philosophical perspective and approach. For this, the philosophies of particular sciences, such as the philosophy of biology, the philosophy of economics, or the philosophy of physics, can provide some support. Moreover, philosophy provides general expertise for the analyses of concepts and the identification of hidden (normative) assumptions, which is crucial for a solid underpinning of interdisciplinary work. Transdisciplinary integration requires an understanding of the specifics of scientific perspectives, approaches, and knowledge on the one hand, and other forms of experience, recognition, and knowledge on the other. Also, it requires a discussion of the relationship between scientific and nonscientific perspectives and types of knowledge. The history of philosophy offers important approaches and insights that can support such a discussion. Of particular relevance would be phenomenology, epistemology, and philosophy of science.

C.U. Becker, *Sustainability Ethics and Sustainability Research*,
DOI 10.1007/978-94-007-2285-9_14, © Springer Science+Business Media B.V. 2012

Philosophy, therefore, should be considered as an *integral part* of sustainability research (Fig. 14.1). It can assist in providing the fundaments for sustainability research. The analysis of science, its role as a meta-structure, its basic assumptions and evaluations, and its reorientation toward a sustainable future all require philosophical reflection. The ability of science to analyze the continuance abilities of certain systems can only be made fruitful for the issue of sustainability if understood as embedded in the aspects of orientation and relationships. The full consideration of this embedment is to an important extent a philosophical task.

There are two main operational ways in which philosophy can contribute to sustainability research. One is by direct input of expertise, i.e., by the participation of philosophers in the inter- and transdisciplinary process of sustainability research. Philosophers can bring in their expertise and knowledge in analyzing basic concepts, specifics of the scientific perspective and particular sciences, methodological presumptions, normative assumptions, etc. They can participate in the process of integrating concepts and methods. They can, in particular, moderate the process of inter- and transdisciplinary integration and discussion, and, for instance, hint at the incompatibility or ambiguity of concepts, hidden assumptions, implicit normative and evaluative elements, and so on. However, as I have argued in Chaps. 12 and 13, sustainability research cannot be organized as a simple division of labor between disciplines. This also holds for the contribution of philosophy. Philosophy as a discipline may contribute its expertise and can support sustainability research in this way, but at the same time it would be crucial that all researchers involved develop the openness to and ability of philosophical reflection on their own disciplines and methods. Only then is a constructive dialogue with fruitful input by philosophers possible, and a substantial process toward an inter- and transdisciplinary sustainability research can develop.

Therefore, the other way in which philosophy can be involved in sustainability research is by contribution to the education and training of researchers. Sustainability research requires demanding cognitive capabilities and communication skills, as well as the ability to reflect on methodological issues and to distinguish and approach normative aspects of sustainability. These skills require training, and this can be done with a limited and passable effort within the education process at universities by the assistance of philosophy—for instance, by providing courses on the philosophy of science, history of science, critical thinking, epistemology, ethics, and communication within the study programs of the sciences. By this, graduate students and prospective researchers will be trained in the ability to reflect on the foundations and preconditions of science as a whole and to reflect on the specifics of and differences between scientific disciplines in particular. They will be trained in the recognition of, and dealing with, normative assumptions. Also, against this background, they will be trained in the adequate communication with politics and societal groups, possibilities of common problem definition, and possibilities of transfer of knowledge between science and society. Furthermore, philosophy may help individuals to consider their own self-identities, particularly their roles as researchers and sustainable persons in the context of societies and the sustainability relations.

Philosophy generally has a crucial meaning for inter- and transdisciplinary teaching. Indeed, inter- and transdisciplinary teaching is mainly philosophical teaching. Of course, one pillar of inter- and transdisciplinary education is disciplinary education. This, however, can simply be done by traditional disciplinary teaching as it exists now. Another pillar is education in the basics of other disciplines. This, too, can be done through focused teaching of the other disciplines in the framework of existing teaching structures and programs. However, the specific characteristic of inter- and transdisciplinary education is the training of capabilities for the integration and transfer of knowledge, the critical reflection of science, and related ethical reflection. This includes, for example, the ability to reflect on the specifics of different disciplines, their self-identities, basic assumptions, concepts, methodological foundations, and inherent normative elements, among others. The training of these capabilities is the domain of philosophy.

However, one has to be aware that developing a new inter- and transdisciplinary sustainability research is also a challenge for philosophy itself. Philosophy will only become fruitful for sustainability research in the full sense described above if it explicitly devotes itself to such a project. This will require some changes and developments of philosophy as an academic discipline itself.

First, philosophy has to recognize its potential for the analysis of crucial societal and global issues and to be willing to contribute to their solution. Philosophers must be willing to make philosophical reflection and the history of philosophical thought fruitful for the discussion of current issues, such as environmental, economic, technological, global, and sustainability issues. Philosophers cannot restrict themselves to a mere discussion of philosophical themes and the history of philosophical thought, but have to consider the relevance of philosophical thought for current issues, and work toward ways to make philosophy fruitful for their discussion.

Second, with this, philosophy has to be aware that the issues we are facing today show new specific characteristics and raise new philosophical questions. Simply applying traditional approaches and concepts may be inappropriate to these issues and may miss the actual philosophical challenges they raise. Philosophers need to carefully reflect on the specifics of current issues and potentially modify traditional philosophical concepts and approaches, or even develop new concepts and approaches. This requires an intense involvement in current academic and public discussion and learning about recent insights of sciences and recent developments in fields of practical action. This is an increasingly demanding task due to the increasing complexity of modern science and fields of action.

Third, a main challenge for the discipline of philosophy will be to maintain a tension between autonomy and historical grounding of philosophical thought on the one hand, and close and careful attention to the present and its specifics on the other. It is this tension that constitutes the particular potential of philosophy for critical reflection and substantial considerations on current conditions and issues. This tension avoids both philosophy becoming unable to contribute to discussions of present issues and philosophy becoming a mere uncritical servant of society or science—a servant who provides simple consulting advice on problems that society or science recognizes and formulates, without analyzing and reflecting on their underlying patterns and assumptions. Neither philosophy nor sustainability research would become fruitful for society and humankind to the full extent by just consulting on given problems. Rather, both also should make problems—that means they should critically reflect on the current condition of society, and indicate and formulate the issues of society and its deeper causes in an independent way.

In conclusion, philosophy should play a crucial role for sustainability research. It should provide the abilities of critical reflection necessary to design and conduct this type of research and play an important role in the training and education of sustainability researchers. Philosophy as an academic discipline should actively involve itself in the task of developing an inter- and transdisciplinary sustainability research and develop the appropriate philosophical tools for this project.

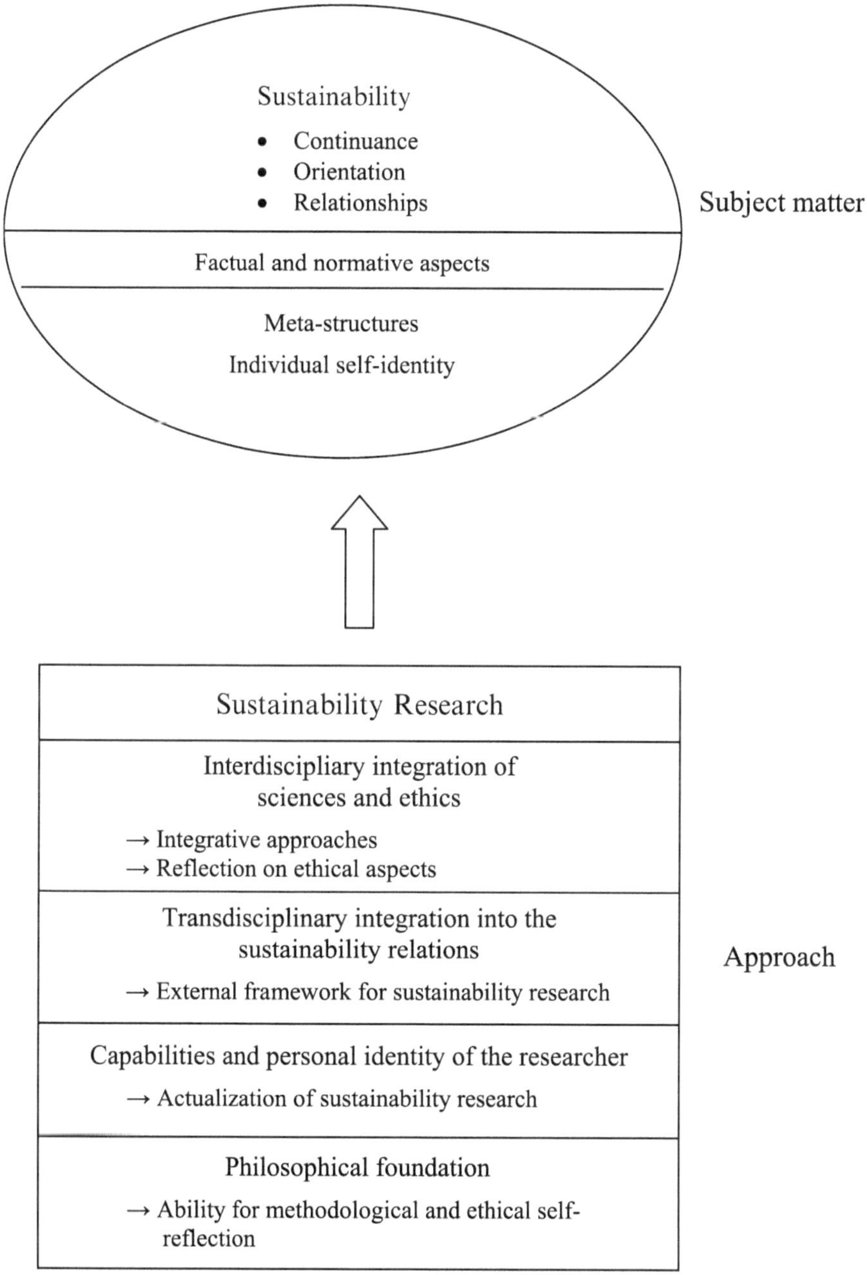

Fig. 14.1 Sustainability research: approach and subject matter

Chapter 15
Conclusion

In this book, I identified the inherent ethical dimension of sustainability, developed an approach of sustainability ethics to analyze this dimension in detail, and discussed the implications of the ethical dimension for sustainability research. The development of my approach of sustainability ethics and the formulation of requirements for sustainability research has been driven by the modern concept of sustainability, its meaning and the issues it implies. The aim has been to develop an ethical and an academic approach appropriate to sustainability. Based on the specific characteristics of the ethical dimension of sustainability, I proposed a new approach of sustainability ethics that is a type of relational ethics integrating both individual and systemic aspects. Further, I argued for a new type of sustainability research that is appropriate to sustainability issues and can particularly refer to the inherent ethical dimension of sustainability. For this, I suggested a new type of self-reflective inter- and transdisciplinary sustainability research that integrates ethics and philosophy.

Sustainability remains a fruitful and important concept for public and academic discussions. The concept is well introduced in a broad range of discourses around the world and has the potential to support a broad, intercultural discussion about the future development of societies. This potential has not yet been fully acknowledged and realized. The discussion about sustainability still needs to be developed. The discussion needs to refer to and include the full meaning of the concept and the fundamental challenges it denotes. This in particular means to refer to the ethical dimension of sustainability and to adequately address and include this dimension in public and academic discourses.

Sustainability is not just about the ability to maintain certain systems. The concept entails a fundamental philosophical issue—the issue of the self-identity of the modern individual and modern (western) societies. Sustainability puts into question the ideal of the human being as an independent and autonomous individual, which has been the underlying ideal of modernity for about the last 300 years. It puts into question the ideal of independence and autonomy from nature, and the ideal of rational control of nature and future, which drove our scientific, technological, and economic development the last centuries with great success. The term sustainability

DOI 10.1007/978-94-007-2285-9_15,

addresses the concern that this success story is close to its end and that the ideals it was based on are one-sided and will not be fruitful guiding principles for the future.

The modern concept of sustainability reminds us that we are also fundamentally related and dependent beings and that we need to take this aspect of human existence more seriously for future actions and life. From an ethical perspective this reminder need not be considered a restriction of the human being but opens the perspective toward its full potential and provides a new ideal of future living. From a virtue ethics perspective, sustainability asks us to develop ourselves as beings fundamentally existing in the threefold relationship with contemporaries, nature, and future in an integrated and excellent way. We are asked to recognize and realize this relatedness of human existence not as restriction but as potential of human excellence. We are asked not to try to overcome our dependence on nature or to try to control nature and the future, but rather to develop ourselves as timely, naturally, and socially contingent beings. We are asked to develop a new attentiveness and respect toward nature, other contemporaries, and future generations, and to develop all three relations in an integrated and excellent way. Sustainability, therefore, means a paradigm shift from an ideal of human independence and rational control toward an ideal of human relatedness and reasonable relationship.

With this, sustainability is a demanding challenge. It is a challenge for the individual person to recognize and understand herself as a sustainable person and to actively develop herself and the sustainability relations accordingly. It is a challenge for societies and the global community, which must change societal and global systems adequately to the ideal of sustainable persons. It is a challenge for academia, which has to redefine its own role for analyzing sustainability issues and develop a new type of self-reflective inter- and transdisciplinary sustainability research. Finally, sustainability is a challenge and obligation for philosophy, which must understand its relevance and potential for the analysis and solution of sustainability issues, develop appropriate philosophical approaches, and involve itself in the discussions about sustainability in a new way.

At the same time, sustainability is an opportunity and provides a fruitful framework for considering and meeting the challenges of the twenty-first century. Sustainability provides an opportunity for the individual to reconsider her self-identity and to recognize the potential of her own being and excellence in the context of the sustainability relations. Sustainability provides an opportunity for societies and the global community to reconsider important systems and their underlying patterns of thought and action, and sustainability provides an orientational framework for the development of a future economy, science, and technology. Sustainability provides an opportunity for science and academia to reconsider their role and meaning for societies and crucial societal issues, and to reorganize and re-orient research in regard to sustainability issues. Finally, sustainability is particularly an opportunity for philosophy to redefine its role for societal discussions and societal development, as well as its role within academic research. Sustainability provides an opportunity for philosophy to recognize and realize its potential for contributing to the analysis and solution of crucial societal issues, to involve itself in academic and scientific research and in the development of societies toward a sustainable future.

In this sense, the modern concept of sustainability is an important and fruitful concept and has the potential to orient the individual, societies, global community, and academic research in their development and improvement. To fully realize this potential requires both theoretical and practical steps. Theoretically, it requires the proper understanding of the meaning of sustainability and the characteristics of sustainability issues, and the development of adequate approaches for their analysis. Practically, it requires concrete and engaged steps to realize these theoretical insights. For this, we need a process of mutual development of individual self-identity and morality, sustainability relations, and societal and global systems. Sustainability is not about a big solution or change in one single step. It is about many small steps of continuous development and change over time.

Subject Index

C.U. Becker, *Sustainability Ethics and Sustainability Research*,
DOI 10.1007/978-94-007-2285-9, © Springer Science+Business Media B.V. 2012

MIX
Papier aus verantwortungsvollen Quellen
Paper from responsible sources
FSC® C105338

If you have any concerns about our products,
you can contact us on
ProductSafety@springernature.com

In case Publisher is established outside the EU,
the EU authorized representative is:
Springer Nature Customer Service Center GmbH
Europaplatz 3, 69115 Heidelberg, Germany

Printed by Libri Plureos GmbH
in Hamburg, Germany